OCEANOGRAPHY AND MINE WARFARE

I0891843

Ocean Studies Board
Commission on Geosciences, Environment, and Resources
National Research Council

NATIONAL ACADEMY PRESS
Washington, D.C.

NATIONAL ACADEMY PRESS • 2101 Constitution Avenue, N.W. • Washington, DC 20418

NOTICE: The project that is the subject of this report was approved by the Governing Board of the National Research Council, whose members are drawn from the councils of the National Academy of Sciences, the National Academy of Engineering, and the Institute of Medicine. The members of the committee responsible for the report were chosen for their special competencies and with regard for appropriate balance.

This report and the committee were supported by a grant from the Office of Naval Research and the Oceanographer of the Navy. The views expressed herein are those of the authors and do not necessarily reflect the views of the sponsors.

International Standard Book Number 0-309-06798-7

Additional copies are available from the National Academy Press, 2101 Constitution Avenue, NW, Lockbox 285, Washington, DC 20055; (800) 624-6242 or (202) 334-3313 (in the Washington metropolitan area); Internet: http://www.nap.edu.

Printed in the United States of America.

THE NATIONAL ACADEMIES

National Academy of Sciences
National Academy of Engineering
Institute of Medicine
National Research Council

The **National Academy of Sciences** is a private, nonprofit, self-perpetuating society of distinguished scholars engaged in scientific and engineering research, dedicated to the furtherance of science and technology and to their use for the general welfare. Upon the authority of the charter granted to it by the Congress in 1863, the Academy has a mandate that requires it to advise the federal government on scientific and technical matters. Dr. Bruce M. Alberts is president of the National Academy of Sciences.

The **National Academy of Engineering** was established in 1964, under the charter of the National Academy of Sciences, as a parallel organization of outstanding engineers. It is autonomous in its administration and in the selection of its members, sharing with the National Academy of Sciences the responsibility for advising the federal government. The National Academy of Engineering also sponsors engineering programs aimed at meeting national needs, encourages education and research, and recognizes the superior achievements of engineers. Dr. William A. Wulf is president of the National Academy of Engineering.

The **Institute of Medicine** was established in 1970 by the National Academy of Sciences to secure the services of eminent members of appropriate professions in the examination of policy matters pertaining to the health of the public. The Institute acts under the responsibility given to the National Academy of Sciences by its congressional charter to be an adviser to the federal government and, upon its own initiative, to identify issues of medical care, research, and education. Dr. Kenneth I. Shine is president of the Institute of Medicine.

The **National Research Council** was organized by the National Academy of Sciences in 1916 to associate the broad community of science and technology with the Academy's purposes of furthering knowledge and advising the federal government. Functioning in accordance with general policies determined by the Academy, the Council has become the principal operating agency of both the National Academy of Sciences and the National Academy of Engineering in providing services to the government, the public, and the scientific and engineering communities. The Council is administered jointly by both Academies and the Institute of Medicine. Dr. Bruce M. Alberts and Dr. William A. Wulf are chairman and vice chairman, respectively, of the National Research Council.

SYMPOSIUM STEERING COMMITTEE

DAVID BRADLEY, *Chair,* Pennsylvania State University, State College
TONY CLARK, North Carolina State University, Raleigh
ROBERT HOLMAN, Oregon State University, Corvallis
MARY JANE PERRY, University of Maine, Orono

Federal Agency Liaisons

MELBOURNE BRISCOE, Office of Naval Research, Arlington, Virginia
LINDA GLOVER, Office of the Oceanographer of the Navy, Washington, D.C.
KATHY SHANEBROOK, Chief of Naval Operations, Washington, D.C.

NRC Staff

ALEXANDRA ISERN, Study Director
DAN WALKER, Symposium Manager and Study Director
SHARI MAGUIRE, Research Assistant
JENNIFER WRIGHT, Senior Project Assistant

COMMISSION ON GEOSCIENCES, ENVIRONMENT, AND RESOURCES

GEORGE M. HORNBERGER, *Chair*, University of Virginia, Charlottesville
RICHARD A. CONWAY, Union Carbide Corporation (retired), S. Charleston, West Virginia
THOMAS E. GRAEDEL, Yale University, New Haven, Connecticut
THOMAS J. GRAFF, Environmental Defense Fund, Oakland, California
EUGENIA KALNAY, University of Oklahoma, Norman
DEBRA KNOPMAN, Progressive Policy Institute, Washington, D.C.
KAI N. LEE, Williams College, Williamstown, Massachusetts
BRAD MOONEY, J. Brad Mooney Associates, Ltd., Arlington, Virginia
HUGH C. MORRIS, El Dorado Gold Corporation, Vancouver, British Columbia
H. RONALD PULLIAM, University of Georgia, Athens
MILTON RUSSELL, Joint Institute for Energy and Environment and University of Tennessee (Emeritus), Knoxville
THOMAS C. SCHELLING, University of Maryland, College Park
ANDREW R. SOLOW, Woods Hole Oceanographic Institution, Massachusetts
VICTORIA J. TSCHINKEL, Landers and Parsons, Tallahassee, Florida
E-AN ZEN, University of Maryland, College Park
MARY LOU ZOBACK, U.S. Geological Survey, Menlo Park, California

Staff

ROBERT M. HAMILTON, Executive Director
GREGORY H. SYMMES, Associate Executive Director
JEANETTE SPOON, Administrative and Financial Officer
DAVID FEARY, Scientific Reports Officer
SANDI FITZPATRICK, Administrative Associate
MARQUITA SMITH, Administrative Assistant/Technology Analyst

Preface

Naval mines present significant challenges for the operational Navy. Between 1950 and 1994, ship casualties caused by mines greatly outnumbered those from all other sources. Mines can essentially halt operations in shallow water, are easily deployed, and are relatively inexpensive to produce. As such, sea mines are rapidly becoming the weapon of choice for many developing countries. Since 1986 the number of mining-capable nations has greatly increased. These increases, along with the low cost and historical success of naval mines, guarantee that these weapons will continue to challenge U.S. Naval forces in the future. As a result, it is important that the warfighter has the tools and training needed to control the battlefield.

The majority of mine warfare occurs in nearshore environments, so an accurate, in-depth knowledge of the oceanographic setting is essential for mission planning and battlefield preparation. As coastal environments are characterized by short-term spatial and temporal variability, the capacity of the warfighter to control the battlefield and ensure mission success is dependent on the ability to understand local oceanography. This ability is greatly enhanced by continued dialogue between the academic community and naval operators to ensure research advances meet the needs of the mine warfighter. Recently, there has been rapid progress in techniques for oceanographic data collection and interpretation. Incorporation of these and future research developments into mine warfare operations will help guarantee the mine warfare community's ability to dominate nearshore operations.

Kenneth H. Brink
Ocean Studies Board, *Chair*

Acknowledgments

The highly successful Symposium on Oceanography for Mine Warfare was a collaborative effort of many individuals from both the Navy and academia. Thus, the content of this report reflects input from a number of individuals, not just the steering committee. In particular, the steering committee would like to acknowledge the assistance of those persons who led working group discussions: R. Beach and L. Frailey (surf zone), A. Fuller (very shallow water), and C. Gunderson (deep water) and those persons who recorded session notes: C. Weilert (surf zone), L. Fry (very shallow water), T. Yoder (shallow water), and C. A. Blain (deep water). The steering committee is also grateful for the assistance of the attendees and experts who prepared background materials that helped to set the stage for later symposium discussions: CDR J. Brown, CDR F. Garcia, LCDR M. Null, and D. Todoroff. The steering committee would also like to recognize the persons involved with the organization and development of the war game: CDR F. Garcia (NO96), CDR T. Hodgson (Naval War College), and CDR M. Shumaker (Office of Naval Intelligence).

In addition to the support and contribution of the many attendees, experts, and individuals listed above, the steering committee would like to acknowledge the efforts of Rear Admiral D. R. Conley, as well as the officers, enlisted personnel, and civilian staff of the Ingelside Mine Warfare Facility and Corpus Christi Naval Air Station, the Office of Naval Research, the Office of the Oceanographer of the Navy, and the Naval Oceanographic Office. Without the help and support of these individuals and groups, this symposium and the resulting report would not have been possible.

For her assistance in data and information gathering, the steering committee would also like to extend its thanks to Constance Carter at the Library of Congress.

This report has been reviewed in draft form by individuals chosen for their diverse perspectives and technical expertise, in accordance with procedures approved by the NRC's Report Review Committee. The purpose of this independent review is to provide candid and critical comments that will assist the institution in making the published report as sound as possible and to ensure that the report meets institutional standards for objectivity, evidence, and responsiveness to the study charge. The review comments and draft manuscript remain confidential to protect the integrity of the deliberative process. We wish to thank the following individuals for their participation in the review of this report: Stephen Boss (University of Arkansas), Peter Howd (University of South Florida), Jules Jaffe (Scripps Institution of Oceanography), Brad Mooney (J. Brad Mooney Associates, Ltd.), Richard Sternberg (University of Washington), and Glen Wheless (Old Dominion University). While the individuals listed above have provided constructive comments and suggestions, it must be emphasized that responsibility for the final content of this report rests entirely with the authoring committee and the institution.

Contents

Executive Summary

Environmental information is important for successful planning and execution of naval operations. A thorough understanding of environmental variability greatly increases the likelihood of mission success. To ensure that naval forces have the most up-to-date capabilities, the Office of Naval Research (ONR) has an extensive environmental research program. This research, to be of greatest use to the warfighter, needs to be directed towards assisting and solving battlefield problems. To increase research community understanding of the operational demands placed on naval operators and to facilitate discussion between these two groups, the National Research Council's (NRC) Ocean Studies Board (OSB), working with ONR and the Office of the Oceanographer of the Navy, convened five previous symposia on tactical oceanography. The sixth and latest symposium in this series, "Oceanography and Mine Warfare," was held in February 1999 in Corpus Christi, Texas.

The symposium, and consequently this report, examined the following issues: (1) how environmental data are used in current mine warfare doctrine, (2) current procedures for in situ collection of data, (3) the present capabilities of the Navy's oceanographic community to provide supporting information for mine warfare operations, and (4) the ability of oceanographic research and technology developments to enhance current mine warfare capabilities. Although, environmental data are also important for offensive mining, a decision was made prior to the symposium and the writing of this report to primarily concentrate on the importance of oceanographic data for mine countermeasures, as it was felt that this community more actively and routinely uses oceanographic data.

Environmental conditions strongly impact mine warfare operations. All aspects of mine warfare, from mine laying to target detection and minehunting and sweeping, are affected by oceanographic and meteorologic variability. To ensure successful mining and counter mining operations in the coastal zone, where most mine warfare takes place, a detailed understanding of the battlefield environment is necessary for all stages of mission planning and execution. In nearshore regions, oceanographic parameters are highly variable. A thorough characterization of these environments therefore requires a combination of historical datasets, in situ sampling, and environmental assimilation models capable of resolving oceanographic variability. This presents a significant challenge to mine warfighters and the community that supports them.

NAVAL OCEANOGRAPHIC CAPABILITIES

The U.S. Navy provides operational oceanographic support for mine warfare through the Office of the Oceanographer of the Navy; Commander, Naval Meteorology and Oceanography Command (CNMOC); the

Naval Oceanographic Office (NAVOCEANO); with research support funded through the science and technology programs managed by ONR. The Oceanographer of the Navy is responsible for collecting and archiving meteorological and oceanographic data. Once collected, the Oceanographer uses databases and interpretive models to evaluate and predict atmospheric and oceanographic processes. Model outputs are then used by the mine warfighter for mission development and planning.

The role of the naval meteorological and oceanographic community has become increasingly important as operations have shifted from more stable ocean conditions characteristic of the open ocean to the rapidly changing conditions of continental shelf and slope environments. NAVOCEANO is currently adapting its ability to collect information in coastal environments, but it is acknowledged that new, creative approaches must be used to supply the environmental data needed to adequately support mine warfare operations. The ability of NAVOCEANO to adjust to more complex environmental conditions has been greatly aided by research and technology efforts supported by ONR.

An additional challenge to the naval oceanographic community results from the shift in operational focus from a dedicated mine warfighting force to a forward-deployed "organic" mine countermeasure (MCM) force. This change has created the need for in situ oceanographic data collection by advance forces such as carrier battle groups, amphibious ready groups, and other naval assets. After collection, these organic MCM forces must be able to rapidly transfer environmental data to NAVOCEANO for processing and conversion into data products that provide the warfighter with knowledge about battlespace environmental conditions.

ROLE OF SCIENTIFIC RESEARCH IN MINE WARFARE

The research community has an important role to play in support of mine warfare operations by developing an increased understanding of nearshore oceanographic processes. This is particularly true for the surf zone (<10 m), which traditionally has been a difficult area for research activities and naval operations.

Advances in coastal zone research will enhance existing quantitative oceanographic models and enable the development of new ones. These models are essential tools for use in nearshore naval operations, as they can be used to forecast coastal oceanographic variability that is likely to impact mine warfare activities and they can also process vast amounts of data and output these data as parameters useful for mine warfare planning and decision making. Furthermore, visualization hardware and software have become increasingly sophisticated. They provide 3D graphical representations of data in an interactive environment where the operator is able to view, as well as manipulate, oceanographic information in an intuitive manner. The use of "virtual environments" changes the way environmental data are used, providing a sense of presence to the user. The MCM community will greatly benefit from the use of such technology.

The academic and engineering communities can also contribute to mine warfare operations in the areas of database development, data transfer, and interactive visualization techniques. Currently, NAVOCEANO and mine countermeasure forces collect extensive high-resolution environmental datasets and acquire additional ones from other nations. Continued developments in management techniques for data archival and retrieval will ensure that these complicated datasets can be used effectively and efficiently. Enhanced data transfer capabilities are also important for mine warfare operations. Presently, a large portion of naval oceanographic data interpretation is performed at NAVOCEANO. To ensure near real time use of in situ environmental data, it is important that wireless data transmissions are available to move data rapidly from ship to shore and back.

With increased use of complex oceanographic datasets it is essential that mine warfare operators receive advanced training enabling them to understand oceanic processes and operate the tools used to interpret oceanographic data. The academic community can play an important role by developing advanced techniques to increase the quality and efficiency of oceanographic training.

EXPANDING CURRENT ENVIRONMENTAL CAPABILITIES FOR MINE WARFARE

At the symposium, mine warfare operators, naval scientists, academic scientists, and representatives of organizations making up the Navy's meteorological and oceanographic infrastructure met in plenary and working group sessions. These discussions brought together individuals with diverse backgrounds and interests to examine the complexities of mine warfare operations in coastal environments. Discussions concentrated on applying knowledge of oceanographic processes to problem solving for mine warfare operations. The steering committee was not charged with determining implementation strategies for recommendations made to the sponsors and the research community during symposium discussions. The steering committee did however identify a number of salient points raised during symposium and steering committee discussions:

- It is important to ensure that naval operational staff have the training necessary to allow them to turn environmental information collected by NAVOCEANO and mine countermeasure (MCM) forces into knowledge the warfighter can use in the battlefield. A greater understanding of coastal zone processes will allow decisionmakers to identify the most advantageous mix of in situ data collection and environmental prediction.
- Developing the ability to quantitatively model and predict environmental variability is an important challenge facing nearshore operations. These models will provide a critical capability for mine warfare mission planning. Enhancement of shallow water modeling is dependent on continued support of coastal zone research aimed at understanding nearshore oceanography.
- The increasing variety of battlefield environmental data being collected and advances in database capabilities have made it apparent that current mine warfare doctrine should be continually adapted, evaluated, and altered to account for parameters other than sediment characteristics. Updated doctrine should be driven by the high-quality environmental data currently being collected by NAVOCEANO and mine countermeasure forces and should account for statistical properties of those data.
- The change in naval focus from deep water to nearshore operations will require the meteorological and oceanographic (METOC) community to continue developing high-resolution environmental data collection capabilities for use in rapidly changing shallow water environments. It is also important that NAVOCEANO continue to enhance database capabilities to store and process oceanographic data from the coastal zone.

1

Introduction

An accurate, in-depth knowledge of the oceanic environment is essential for planning and execution of naval operations. Techniques for rapid oceanographic data collection, assimilation, and dissemination either directly or remotely are developing rapidly. In addition, quantitative models incorporating these data for prediction of environmental variability are continually being developed and refined. The Office of Naval Research (ONR) has provided extensive funding for academic oceanographic research to ensure that naval operators have access to state-of-the-art oceanographic data collection methods and interpretive models. To improve the academic ocean science community's understanding of operational demands placed on naval units and the oceanographic needs of naval operations, the Ocean Studies Board (OSB) of the National Research Council (NRC), working with ONR and the Office of the Oceanographer of the Navy, convened five previous symposia on tactical oceanography (NRC 1991, 1992, 1994, 1996a,b, 1997). These symposia have proven to be an important forum for facilitating dialogue between academic scientists and naval warfighters, helping academic researchers identify potential research areas of maximum value to the Navy and enabling them to make efficient use of naval research funds.

Mine warfare operational decisions are influenced by local oceanic conditions; thus, accurate characterization of environmental parameters is important for successful battlefield operations. The collection of high-resolution meteorological and oceanographic datasets to accurately describe the battlefield is essential for establishing "realistic" mine clearance time lines and ultimately "realistic" estimates of risk to maritime forces. A thorough understanding of the coastal water column, the nature of the coastline, seafloor variability and rigidity, sub-seafloor characteristics, and the concentration of biological growth on or near the seafloor can help ensure mission success.

The sixth OSB symposium on tactical oceanography, "Oceanography and Mine Warfare," was designed to examine the state of environmental knowledge and predictive capability relevant to mine warfare operations and to identify areas of ocean science that will enhance the Navy's mine warfare capability today and for the next 20 years. Emphasis was placed on practical demonstrations and interactive discussions to determine areas of shared interest for naval warfighters and academic scientists. A major result of this interaction was the identification of oceanographic science and technology advances that can be exploited by the Navy to improve its mine warfare capabilities.

The symposium included a discussion of oceanic processes (e.g., tides, currents, and wave action) that control the transport of momentum and material in the environment between the 500-ft. (150 m) depth contour and the seasonal high-tide mark. Specifically, discussions addressed the dynamic relationships between fluid, sediment, and biota and the resulting effects on boundary stability, water column visibility, and seawater and sediment acoustic properties. Participants reviewed present capabilities to monitor and predict physical parameters, such as mean flow

BOX 1-1 Statement of Task

This symposium was designed to bring together members of the academic community, the U.S. Navy (fleet operators, meteorologists, and oceanographers), and scientists, technologists, and managers from naval warfare centers and defense programs to:

- address timely operational problems and fleet mission needs in which meteorological and oceanographic research and development (R&D) play a role;
- enhance communication among the basic and applied research communities, as well as with naval forces engaged in mine warfare; and
- enable an extended group of researchers to become familiar with challenging naval issues related to the use of environmental information in mine warfare.

dispersal, instantaneous maximum forces in the fluid and on the seabed, and other environmental factors important to conducting mine warfare operations in the tactically important nearshore environment. The effects of the marine boundary layer and water column visibility on remote-sensing capabilities and operations were also discussed.

The NRC appointed a symposium steering committee consisting of one OSB member and three ocean scientists with varied expertise (Appendix A). The steering committee was tasked to work with Navy personnel to identify topics to be covered, determine the most appropriate speakers to address the symposium, name members of the academic community who should be encouraged to attend the symposium, and write this unclassified report (Box 1-1).

THE IMPACT OF OCEANOGRAPHY ON MINE WARFARE

> Understanding the oceans is fundamental to our national security, as well as to global economic and environmental well-being. A robust competency in oceanography is a core requirement of the U.S. Navy. It is so vital to the success of naval operations that the Navy must lead in focusing national attention on ocean policy and programs.
>
> —Admiral Boorda, Chief of Naval Operations, 1995

Oceanography is a core competency of the U.S. Navy. Oceanographic data are acquired using unmanned aerial and underwater vehicles; sensors on ships, airplanes, and submarines; expendable sensors; and remote sensors, such as satellites. Collection of these data helps to ensure mission success and is of fundamental importance to national security. As such, the Navy aggressively maintains oceanographic data collection capabilities, many of which reside with the Naval Oceanographic Office (NAVOCEANO).

In littoral[1] warfare, strategic, operational, and tactical mobility are obvious advantages to naval forces that rely on unobstructed sea lanes. Naval mines can diminish or deny these advantages by reducing freedom of movement and preventing naval forces from controlling the battlefield. Naval mines not only delay offensive maneuvers, they also provide an adversary with the time needed to shape the battlefield and move forces into more advantageous locations. Since mines are plentiful and relatively inexpensive, they are an obvious and effective warfighting alternative for cash-poor developing nations. This was exemplified during the Gulf War, where mine damage to three U.S. warships (USS *Samuel B Roberts*, *Tripoli*, and *Princeton*) was in excess of $125 million, whereas the mines that caused the damage, including two of World War I vintage, cost approximately $30,000 (Boorda, 1999). In fact, most of the U.S. ship casualties from 1950 to 1994 were from naval mines (Figure 1-1).

Since the end of the Cold War, there has been a shift in naval operational focus from the open ocean to nearshore environments (Table 1-1). As most mine warfare operations occur in coastal areas, the potential for mines to frustrate

[1]Littoral, *n.* 1. The designation of the shore area between tide marks. 2. The benthic zone extending from the highwater point on the beach to the edge of the continental shelf (Lapidus, 1987). The naval definition is more specific and goes on to state that the littoral are the regions relating to or existing on a shore or coastal region, within direct control of and vulnerable to the striking power of naval forces.

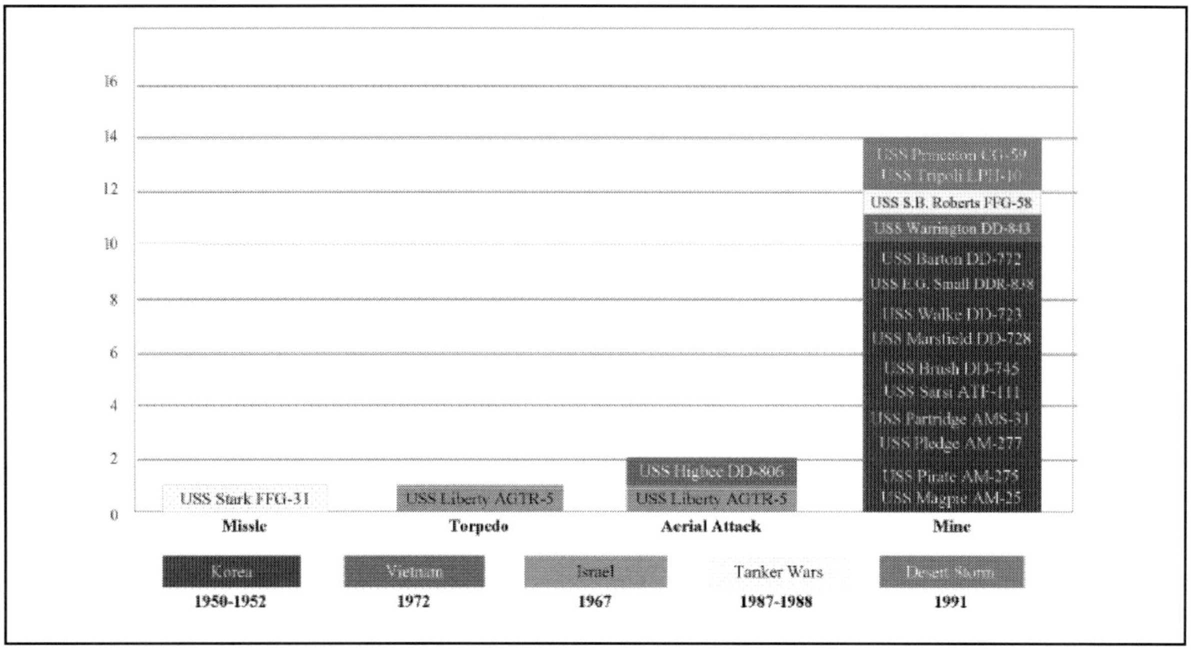

FIGURE 1-1 U.S. ship casualties due to missiles, torpedoes, aerial attack, and mines during military conflicts from 1950 to 1994 (Avery, 1998). No additional ship casualties have occurred since 1994.

U.S. Naval operations has greatly increased (Box 1-2). This operational change has meant that the Navy needs effective mine countermeasures. The success of these countermeasures depends greatly on the capability to determine and accurately predict environmental parameters.

In no other phase of littoral warfare do environmental considerations in both tactics and planning play a more dominant role than in mine warfare. Virtually every environmental parameter in the dynamic nearshore environment influences military operations. This provides a great challenge to both the warfighters and the meteorological and oceanographic (METOC) community that supports them. Thus, for nearshore operations, the Navy's need to respond to environmental variability is amplified.

Another complication for operations in coastal environments is that access to these areas is often either limited or challenged. This restricts the ability to monitor rapidly changing environmental conditions in these areas, which in turn hinders the development of coastal zone environmental databases.

Probably the greatest challenge currently facing the METOC community is the need to develop quantitative predictive models of nearshore oceanographic processes to turn vast amounts of environmental data into knowledge about the battlefield. Outputs of these models can then be incorporated into tactical decision aids to assist the mine

TABLE 1-1 Differences in Operational Parameters Between the Cold War and Today

During Cold War	Today
– USSR/Warsaw Pact	– Multiple threats
– Global warfare	– Regional warfare
– Ocean basins	– Littoral and hinterlands
– Deep water	– Expeditionary warfare
	– Mine
	– Amphibious
	– Special forces

BOX 1-2 Significance of the Coastal Environment

- 95% of the world's population lives within 600 miles of the sea.
- 80% of all countries border the coast.
- 80% of world's capitals lie within 300 miles of a shoreline.

Source: Ellis, 1998

warfighter in mission planning and development. The academic community has an important role to play in providing fundamental research on the dynamics of oceanographic processes. This knowledge will provide the foundation for model developments and enhancements.

> The tactical advantage will probably depend not on who has the most expensive, sophisticated platforms—but rather on who can most fully exploit the natural advantages gained by a thorough understanding of the physical environment.
>
> —Rear Adm. W. G. "Jerry" Ellis, U.S. Navy, Oceanographer of the Navy, 1999

SETTING AND DESCRIPTION OF THE SYMPOSIUM

To encourage interactions between mine warfighters and academic scientists, the "Oceanography and Mine Warfare" symposium was held at the Naval Air Station, Corpus Christi, Texas on February 9-11, 1999 (Appendixes B and C). Prior to the symposium, attendees were organized into four groups matching the depth divisions defined for naval operations (Figure 1-2): surf zone (0-10 ft.), very shallow water (10-40 ft.), shallow water (40-200 ft.), and deep water (> 200 ft.). Day one of the meeting emphasized contacts among attendees, personnel of the Mine Warfare Command (COMINEWARCOM), and personnel of the Mine Warfare Training Center (MWTC). Participants had a productive day touring the CH-53 airborne mine countermeasure (MCM) helicopters and talking with crew members at the Naval Air Station in Corpus Christi. At Ingleside, there were tours of the MCM flagship USS *Inchon* (MCS-12), USS *Warrior* (MCM-10), USS *Falcon* (MHC-59), the Helicopter Mine Countermeasures Squadron (HM-15), and the MCM training facility. In every case, there were excellent briefings with time for questions, discussion of equipment on-board the various vessels, and a chance to get a feel for the operational conditions faced by the U.S. Navy. This face-to-face interaction between the operators and the scientific community was a critical element of the symposium.

Day two of the symposium introduced the attendees to the typical mission needs for mine warfare operations. Groups participated in a single-sided, seminar-style war game (Appendix D). The war game complemented interactions with the operators on the first day of the symposium and served to focus discussion toward environmental understanding and prediction, and away from the intermediate issues of operators. This change in focus led to discussions on tactical exploitation and operational limitations resulting from environmental influences and set the stage for the remainder of the symposium.

The final day of the symposium focused on lessons learned and on identifying ways for the ocean science community to support mine warfare. Attendees were again organized into their respective working groups to participate in a problem-solving dialogue. By the end of the symposium, participants highlighted research areas where advances can provide operational capabilities essential to COMINEWARCOM activities. Thus, the attendees helped identify future oceanographic and meteorological directions for the U.S. Navy and the Mine Warfare Command.

REPORT STRUCTURE AND SCOPE

The aim of this report is to discuss specific oceanographic requirements of the mine warfighter in the context of more traditional academic ocean sciences and help the academic community better understand the research needs of

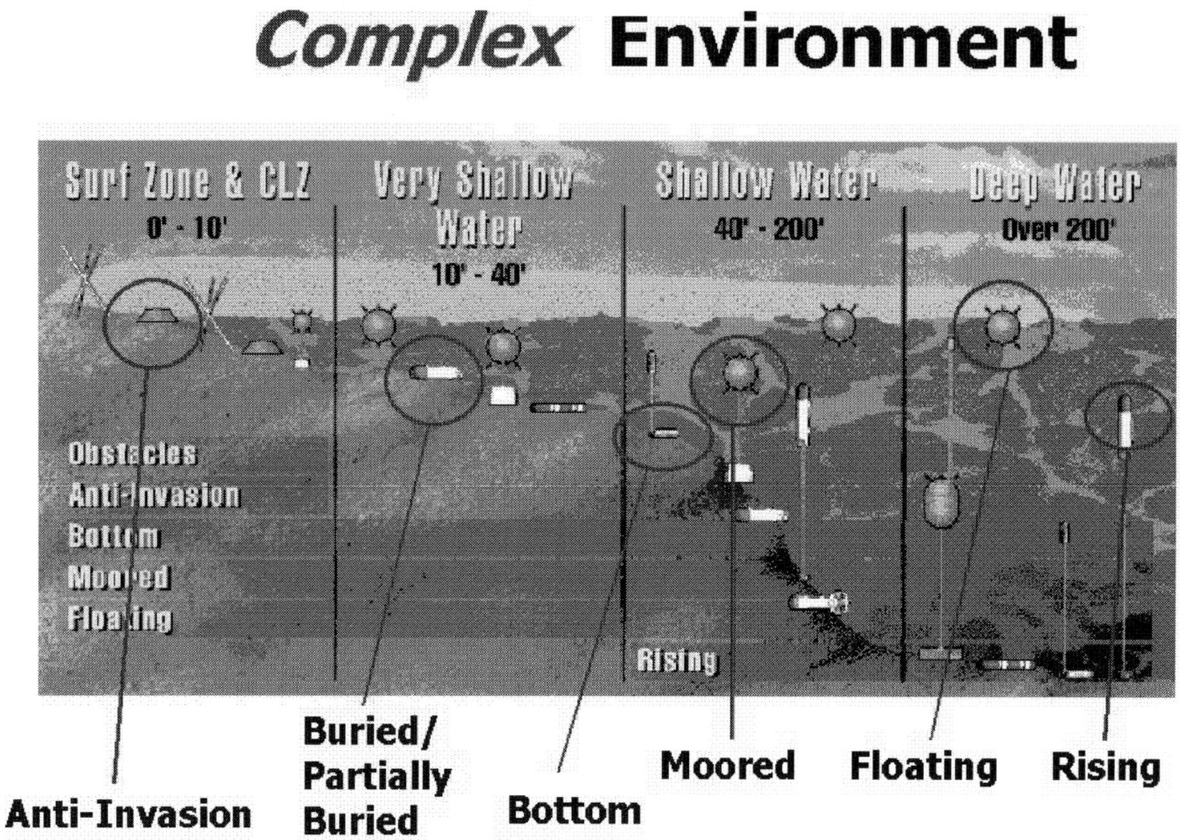

FIGURE 1-2 Major depth divisions for naval operations, with the various mine types common at each depth.

the mine warfare operator. In addition, this report documents the substance of the symposium, thus making the meeting outcomes accessible to a wider audience than the limited number of individuals present at the conference.

Chapter 1 provides background information and justification for the symposium. Chapter 2 provides an overview on mine warfare for the academic scientist unfamiliar with naval mining. Chapter 3 discusses the challenges faced by mine warfare operators as a result of nearshore oceanographic variability, outlines the role of oceanography in mission planning, and summarizes current METOC and ONR support for mine warfare operations. Chapter 4 discusses environmental influences on mine warfare, specific capabilities needed to enhance warfighting, and outlines research issues and solutions. Chapter 5 provides a summary and discussion of recommendations resulting from the symposium and committee discussions.

The information in this report is based on declassified meeting notes and summaries from plenary and working group discussions, supplemented with post-meeting review discussions by members of the steering committee and members of various Navy units. It must be noted that the symposium, and the information contained in this report, primarily concentrated on the importance of oceanographic data for mine countermeasures. Environmental data are also important for offensive mining but, during the organization of the symposium, the statement of task was focused on mine countermeasures as it was understood at the meeting that this community more actively and routinely uses oceanographic data.

2

Mine Warfare: An Overview

The naval mine is an efficient force multiplier and is one of the most cost-effective weapons in the naval arsenal (Table 2-1). Mines are small, easy to conceal, cheap to acquire, require virtually no maintenance, and can be easily and simply laid from almost any type of platform. Naval mines can be used to deny hostile forces access to the coastal zone and to defend important targets, such as ports, anchorages, and offshore structures, from amphibious or seaborne attack. Mines can quickly wipe out, or seriously impair, the effectiveness of surface and submarine forces. Emplaced mines are also difficult to counter and neutralize, especially in the presence of hostile forces. Because of these factors, mines are one of the most effective and deadly weapons that a naval force can employ.

The Navy must be prepared to counter technologically advanced mines, as well as their low-tech predecessors. The success of simple World War I- and World War II-vintage mines means that these weapons will undoubtedly continue to threaten U.S. Naval forces. Since the mid 1980s, the number of naval mines and mining-capable countries has increased dramatically. At present, more than 50 countries, including those in politically sensitive areas, have mining capabilities with more than 300 mine types available. The rapid increase in mining-capable countries results, in part, from a proliferation of mine producers and exporters. Since 1988 there has been a 75% increase in the number of mine-producing countries (Avery, 1998). Many of these producers are manufacturing mines of higher capability thus requiring a parallel increase in the technology for countering these mines. Despite this trend toward more sophisticated mines, the proliferation of older vintage mines still pose a significant threat.

Naval mines provide a great advantage to foreign powers, enabling them to control nearshore operational areas by channeling, blocking, deflecting, disrupting, or delaying opposing forces and preventing them from achieving their objectives. Mines can also jeopardize the steady flow of seaborne materials, equipment, and fuels needed to sustain operations of land-based air and ground forces. As a majority of materials sent to support these forces comes by sea, the ability to close vital waterways provides a significant strategic threat to land-based operations.

Present-day political situations dictate that the military adapt to potential threats that can come from almost any quarter (including terrorism) rather than only from a global superpower. Naval operations are increasingly focused on coastal waters and adjacent land areas rather than the open ocean (Box 2-1). This is a significant change for the Navy, now, in addition to maintaining a strong emphasis on anti-submarine warfare, it is also necessary to make an equally strong effort in the area of expeditionary and mine warfare. In times of conflict, domination of coastal zone operational areas will largely depend on the ability to remove or delineate the emplaced mine threat and prepare the battlefield for follow-on forces in a timely fashion (Box 2-2).

TABLE 2-1 Relative Costs of Common U.S. Mine and Missile Types in 1986

Mark 60 captor mine	**$ 113,000**
Tomahawk cruise missile	**$ 3,100,000**
Harpoon surface-to-air missile	**$ 900,000**

Source: Morison, 1995

An important aspect of offensive mine warfare is that the mere threat of mine deployments can have the same effect as actually establishing a minefield by tying up assets that could otherwise be put to better use. Opposing forces are unlikely to risk personnel and equipment after intelligence indicates mines may be present. In addition, in areas where the seafloor is rough with significant Non-Mine, mine-like Bottom Objects (NOMBO) an adversary can create a highly effective minefield with only a few mines. Dummy mines, or mine-like objects (such as empty barrels), can also be laid in a minefield. These objects can greatly complicate mine countermeasure (MCM) efforts as they must be hunted and classified to ensure that they are not live mines. These situations make it difficult for the mine warfighter to distinguish between real and imagined threats.

Both offensive and defensive naval mine warfare activities (Figure 2-1) are coordinated through Mine Warfare Command (COMINEWARCOM; Figure 2-2). The Naval Mine Warfare Command is accountable for development of naval mining strategy; for overseeing the Navy's inventory of underwater mines; and for directing standing MCM activities, including surface, air, and underwater assets. COMINEWARCOM is responsible for 27 ships, including the USS *Inchon*, the naval MCM command and support ship, 14 open-ocean MCM ships, and 12 coastal minehunters. In addition to these surface assets, during MCM operations, COMINEWARCOM is the operational commander of two helicopter MCM squadrons and explosive ordinance disposal (EOD) detachments. COMINEWARCOM is also

BOX 2-1 Military Significance of Nearshore Areas

- A majority of politically sensitive countries have ocean borders.
- Many crucial shipping lanes are narrow; thus mining is highly effective in these areas.
- Many of the supplies to land-based operations must be transported by sea through the littorals.

BOX 2-2 Key Priorities Established by the Navy to Ensure a High State of Readiness for Future Mine Warfare Capabilities

- Develop a clandestine mine surveillance, reconnaissance, and detection capability;
- Develop an all-source intelligence database to monitor ongoing mine research, development, and technologies throughout the world;
- Develop an organic mine-hunting and neutralization capability in carrier battlegroups and amphibious ready groups so they can conduct mine countermeasure (MCM) operations en route;
- Develop an in-stride mine detection and neutralization capability;
- Upgrade MCM system capabilities;
- Develop an affordable mine force command and control, communications, computers and intelligence (C^4I) capability to enable real-time data exchange among mine countermeasure (MCM) units and between senior expeditionary force commanders;
- Integrate mine warfare and amphibious warfare into training, education, gaming and simulation, exercises, and deployments;
- Integrate regular and reserve units for efficient MCM forces; and
- Develop naval mines, doctrine, and tactics for use in offensive and defensive operations.

Source: Morison, 1995

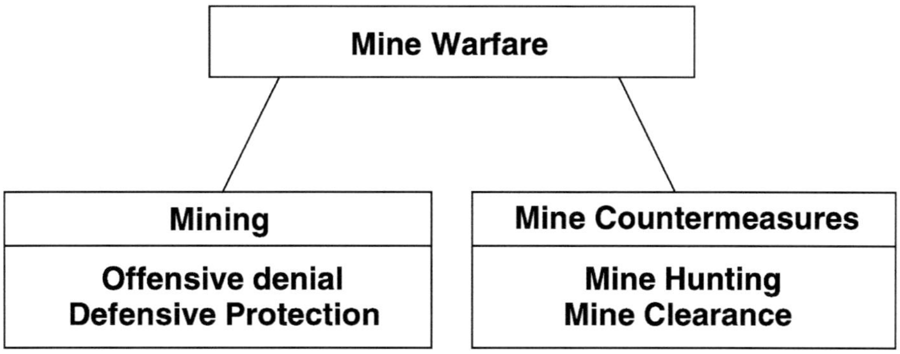

FIGURE 2-1 The two operational subdivisions of naval mine warfare: mining and mine countermeasures (MCM). Mine laying can deny hostile forces access to a strategic location while also protecting home waters. MCM includes the search for and destruction of emplaced mines.

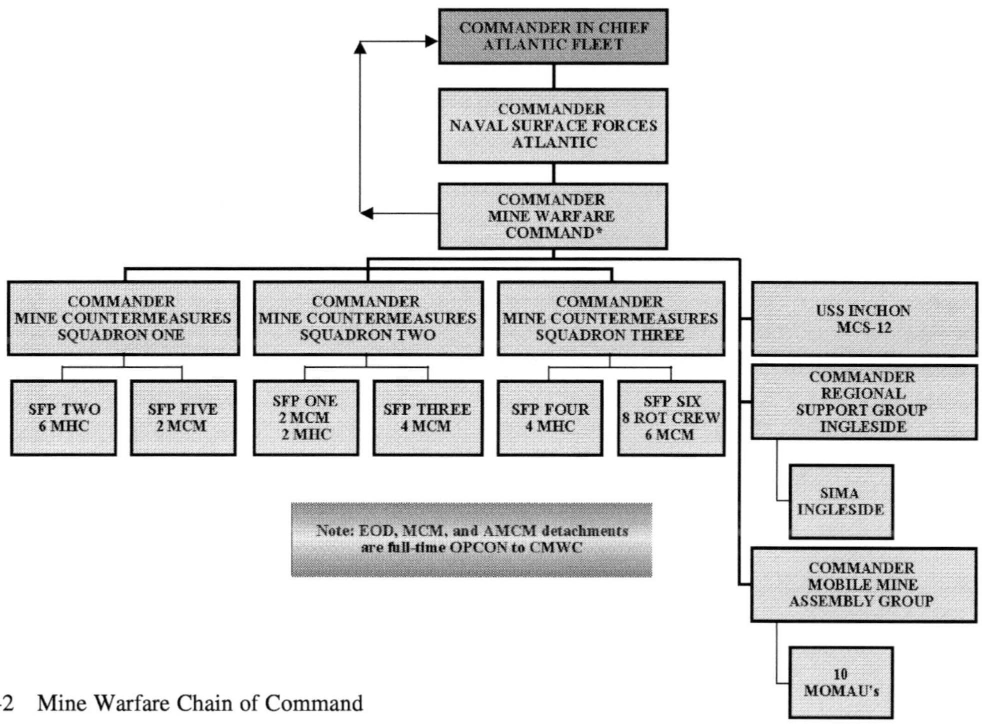

FIGURE 2-2 Mine Warfare Chain of Command

* Commander, Mine Warfare Command reports directly to the Commander in Chief of the Atlantic Fleet.
AMCM – Airborne Mine Countermeasures
CMWC – Commander, Mine Warfare Command (also referred to as COMINEWARCOM)
EOD – Explosive Ordinance Disposal
MCM – Mine Countermeasures
MCS – Mine Countermeasures Command, Control and Support Ship
MHC – Mine-hunter, Coastal ship
MOMAU – Mobile Mine Assembly Unit
OPCON – Operational Control
SFP – Ships Force Package
SIMA – Shore Intermediate Maintenance Activity
ROT – Rotational

responsible for the Mobile Mine Assembly Group (COMOMAG), which includes all Mobile Mine Assembly Units (MOMAUs) and detachments. This group maintains the U.S. Navy's mine stockpile, and when directed, assembles and completes the final preparation of mines to support mining operations.

The Mine Warfare Command mission statement includes the development and evaluation of mine warfare doctrine, tactics, and equipment. In addition, as technical advisor to the fleet commanders in chief, COMINEWARCOM provides mine warfare planning, intelligence on foreign mine capabilities, and develops countermeasures to protect naval assets.

HISTORY OF MINE WARFARE

The first successful sea mine was invented by David Bushnell in 1776 (he is also credited with the construction of the first submarine in 1775). It consisted of a wooden keg filled with gunpowder and fitted with a primitive contact fuse. Even though the Bushnell mines were not very reliable, they were a successful psychological deterrent to British naval operations in harbors and rivers.

Between the American Revolution and the Civil War, there were continual minor improvements to sea mines, but in general, mines were not a priority for the Navy. In the Civil War, Confederate lieutenant Hunter Davidson invented an electrical mine that could be detonated from shore. The Confederates also invented a mine, the "Singer" mine, that detonated on contact. Despite the malfunctions of the Confederate mines, they sank seven Union ships during the war.

The next major mining development was the Hertz horn invented in 1866. This metal horn, variations of which are still in use, contained a solution of a potassium dichromate in a glass tube that broke on impact. Once the solution was released it completed an electrical circuit in the mine battery that fired the mine.

In the Russo-Japanese War (1904-1905), strategic mining by the Japanese eventually led to Russia's defeat. Soon after, sea mines played a major role in World War I, particularly in the Dardanelles-Gallipoli campaign and the North Sea mine barrage. In the Dardanelles Strait, allied forces attempting to reduce Turkish defenses met with mines laid in the strait. Four ships were lost or damaged and plans to take Istanbul were given up. In the North Sea, to limit access to the open Atlantic, the Allies laid 70,117 mines in a 230-mile area in the northern portion of the English Channel. Smaller fields were laid in the southern portion of the channel (Morison, 1995). By the end of World War I, the mine had proven itself to be a highly effective naval weapon that dramatically changed war at sea. But, extensive mine laying during the war necessitated significant advances in MCM. The British invented most of these countermeasures, and some of these techniques are still in use today.

The advent of World War II brought about a new and expanded role for sea mines after a relative period of inactivity between the two world wars. Part of this expanded role was an increase in the use of mines as offensive weapons. In addition, World War II saw the development of mines detonated using acoustic and magnetic influences and the first use of aircraft as minelayers.

During World War II, mines were of critical strategic importance. For example, German mine campaigns in the Baltic Sea effectively sealed the Soviet fleet in port for the entire war. In the Pacific, Allied "Operation Starvation" laid over 11,000 influence mines in Japanese shipping routes to Southeast Asia and also in Japanese harbor entrances (Morison, 1995). This initiative resulted in the damage or destruction of most of the surviving Japanese merchant marine fleet, and it effectively sealed off sea trade to and from Japan. By the end of World War II, sea mines had again proven their effectiveness, and resulted in the sinking of one ship for every 35 mines laid.

Since sea mines provide offensive and defensive capabilities far exceeding their cost, post-World War II naval warfare saw expansions in the use of mines by underdeveloped nations in conflicts with larger superpowers. In the Korean War, offensive mining by communist forces immobilized U.S. Naval operations for more than a week during the landing at Wonsan (Zwolski, 1998). Approximately 3,000 Russian-made contact and magnetic mines caused a fleet of 250 ships to wait off the coast while 10 American minesweepers tried to sweep a clear channel through the minefield (Morison, 1995).

> We have lost command of the sea to a nation without a navy, using weapons that were obsolete in World War I and laid by vessels that were used at the time of the birth of Jesus Christ.
>
> —Rear Adm. Smith, Commander, Amphibious Task Force, Wonsan, Korea, 1950

The most recent example occurred during Operation Desert Storm where mine laying by Iraqi forces in waters off the Kuwaiti coast resulted in extensive damage to three U.S. warships, (USS *Tripoli*, USS *Princeton*, and USS *Samuel B Roberts*; Figure 1-1). Two of the mines that inflicted the damage were World War I vintage.

MINE CLASSIFICATION

Mines are classified according to three characteristics (Box 2-3). The first characteristic is its position in the water column. Bottom mines, which can also include buried mines, are generally found in shallow water, where either surface craft or submarines can easily trigger them. This type of mine is usually influence activated (see below). Moored mines (also called tethered mines) are usually used in deeper water and can be positioned at any depth in the water column. Moored mines are highly effective against both ships and submarines. Tethered mines are primarily used as anti-submarine weapons and can remain passive until the target satisfies its firing criteria. Some tethered mines contain torpedoes or mobile weapons that have the ability to home on targets. Drifting mines move with the prevailing current direction and are designed to float at or just below the sea surface. Drifting mines were banned in 1907 by the Hague Convention, but they are still used by many "rogue" nations. As little as 20 years ago, a minefield could only be emplaced in relatively shallow water (less than 600 ft.). New trends in mine development enable mines to be positioned in deeper water where they will augment anti-submarine and ship barriers and present serious difficulties for minehunting and sweeping.

The second characteristic is its method of delivery (although most mines can be altered for a different mode of delivery). Aircraft-laid mines are generally used for offensive actions where mines need to be rapidly deployed. These mines are dropped like bombs from an aircraft. Most mines in the current U.S. Naval arsenal are aircraft laid mines (Table 2-2). Surface laid mines can be deployed from a variety of ships, enabling them to be laid secretively. The United States no longer maintains a stockpile of surface-delivered mines, although this is a common delivery method for other nations (Table 2-2). Submarine-laid mines are designed to be fired out of torpedo tubes and thus, these types of mines are often torpedo shaped.

The third characteristic is its method of activation. Influence activation is the most common method (Table 2-2). Detectors are used to sense changes in fluid pressure or acoustic, magnetic, and electric fields (or a combination of all four). Sensors can be designed to distinguish between different types of vessels or to have delays before detonating to ensure maximum damage to the triggering vessel (Table 2-2). Contact mines are the oldest technology for mine activation and, in the simplest version, are designed to detonate when the "horn" on the mine is bent. Controlled mines are triggered remotely, using cables connected to shore, although new designs can be remotely controlled, which will effectively allow a minefield to be turned on and off when needed to allow ships to pass through the mined area. This type of mine is typically used defensively and is particularly effective in preventing entrance to straits or confined areas.

There are numerous mines in use by U.S. and foreign navies (Table 2-2). These mines range from simple limpet mines that are magnetically or mechanically attached to an object and detonated with a time fuse (Yugoslavian M-71) to the highly sophisticated Swedish BGM 100 mine, which employs an advanced mine shape and sophisticated sensors to prevent detection. Mines are constructed in a wide range of sizes and can operate in water depths of 0-500 m (Table 2-2). These parameters, when considered along with the high oceanographic variability in shallow water environments, determine the ability of MCM forces to detect mines in the coastal zone.

BOX 2-3 Mine Characteristics

Position in Water	Method of Delivery	Method of Activation
• Bottom	• Aircraft laid	• Influence
• Moored	• Surface laid	• Contact
• Drifting	• Submarine laid	• Controlled

TABLE 2-2 Common Mines in Use by Various Nations

Country	Name	Type	Description
Brazil	MCF-100	Moored contact mine	Can be programmed to remain inert for a fixed period of time, then self-release and anchor itself at the desired depth; can be fitted with an influence sensor.
Denmark	MTP-19	Cable-controlled mine	Fully remote controlled; can be operated at distances of 12 km or more; consists of a portable weapon control unit, distribution box, and the mine itself.
Germany	SM G2	Ground influence mine	Has a non-magnetic casing with acoustic, magnetic, and pressure influence sensors; mine will detonate when preset influence parameters are recognized.
	DM 211 DM 221	Anti-frogman underwater signal charge	Anti-frogman depth charge protects ships and harbor installations against divers; signal charge is used for encoded submarine-to-surface ship communications.
Iraq	Al Kaakaa/16	Floodable submersible mine	Possibly the largest mine in the world that is designed to destroy offshore structures; operates even in very deep water; timer or remote-control detonation.
	Sigeel/400	Ground mine	Seabed mine for deep and shallow water; for use against medium and large targets; can be ship or helicopter deployed.
Italy	MR-80	Seabed influence mine	Actuated by magnetic, acoustic, and pressure influences from the target; body is composed of epoxy resin and glass fibre.
	MP-80	Seabed influence mine	Similar influence activation as the MR-80 but with microprocessors to discriminate between target and countermeasures.
	Murena	Seabed influence mine	Similar to MP-80 but with an advanced microprocessor-firing device.
	Manta Mine	Shallow seabed influence mine	A dual influence (magnetic and acoustic) anti-invasion mine shaped to rest firmly on the seabed even in strong flows.
	Seppia	Moored influence mine	Programmable mine that can operate on any type of bottom against all types of targets; can select targets and discriminate against countermeasures; can be remotely controlled.
Russia *	MDM series (MDM 1-5)	Seabed influence mines	Can be laid from a variety of platforms, activated with either acoustic-magnetic or acoustic-magnetic-pressure sensors.
	SMDM series	Self-propelled seabed mines	Similar to MDM mines but are placed in the body of a torpedo to be laid by submarines.
	PMK-1	Underwater rocket-powered torpedo	A combined mine based on a torpedo with an on-board computer and sensor to identify the target and compute the required trajectory; detonation using combined influence, contact, and time fuse.
	MSHM	Continental shelf mine	Acoustic sensor detects and identifies the target, computes the required trajectory before the underwater rocket is fired to home in on the target.
Spain	MO-90	Moored influence mine	An intelligent multi-influence mine, parameters can be programmed into the weapon's computer before deployment; incorporates anti-minesweeping countermeasures.
	Mila-6B	Naval limpet mine	Time-fused mine that can be used as a demolition charge; attached to underwater structures mechanically or magnetically.

TABLE 2-2 Common Mines in Use by Various Nations

Length (mm)	Width (mm)	Height (mm)	Weight (kg)	Charge (kg)	Operational Depth (m)
1,400	1,020	1,500	770	160	3-50
1,000	1,090	1,128	800	300	3-20
2,000	600	600	750	N/A	N/A
268 DM211 145 DM221	60	60	1.4 DM211 0.8 DM221	0.5 DM211 0.05DM221	6
3,400	3,400	3,000	16,100	9,000	N/A
700-980	700-980	850	535	400	N/A
1,646-2,750	533	533	565-1,035	380-865	5-300
2,096	533	533	780	630	5-300
2,096	533	533	780	630	5-300
980	980	470	225	150 or 180	2.5-100
1,560	533	533	870	200	20-300
1,530-3,060	450-790	450-790	525-1,500	300-1,350	15-35 to 15-300
7,900 SMDM-1 11,000 SMDM-2 7,830	533 SMDM-1 650 SMDM-2 533	533 SMDM-1 650 SMDM-2 533	1,980 SMDM-1 5,500 SMDM-2 1,850	480 SMDM-1 800 SMDM-2 350	4-100 SMDM-1 8-150 SMDM-2 200-400
4,000	533	533	1,500	250	60-300
1,180	1,090	1,690	1,060	300	5-340
350	350	150	65	N/A	0-40

TABLE 2-2 Continued

Country	Name	Type	Description
Sweden	BGM 100 (Rockan)	Anti-invasion ground influence mine	Advanced gliding mine shape allows mine-laying over a wide area while covering the minimal distance; low-profile shape makes it difficult to detect; constructed of reinforced plastic.
	BMM 80	Moored influence mine	Programmed to anchor itself automatically at the desired depth before sinking.
	BGM 601	Ground influence mine (Bunny)	Developed as a submarine weapon to be attached to the outer hull; incorporates multiple sensors with sophisticated logic making it resistant to countermeasures.
	BGM 600	Cable-controlled mine	Developed to provide a rapid means of deploying a controlled minefield.
United Kingdom	Sea Urchin	Programmable influence mine	An intelligent seabed mine that can be programmed to detonate by a range of influences when the target is at the closest approach point.
	Stonefish mines	Ground influence mine	Designed to be modular to fit any requirement; detonation by acoustic, magnetic, and pressure sensors with signal processing capabilities.
United States	Mk 52, 55, 56, and 57	Seabed influence mines	These mines are now obsolete although many are still in inventory; activated using magnetic, acoustic, and pressure sensors.
	Mk 36, 40, 41, and 115A	Air-laid influence mines	Detonated using magnetic and acoustic sensors; intended for use in shallow water; obsolete although many are still in inventory.
	Quickstrike series	Air-launched seabed mine	Group of mines with different cases but common target detection and classification mechanisms; based on conversion of existing ordinance.
	Mk 60	Encapsulated torpedo	A Mk 46 torpedo inserted into a mine casing; used for anti-submarine warfare; can classify targets and initiate the release of the torpedo; employs both passive and active sensors.
	Mk 67 SLMM	Air-launched seabed influence mine	Self-propelled mine for use in shallow water and where covert mining is desirable; detonation using magnetic, acoustic, and pressure sensors.
Yugoslavia	M66	Diversionary underwater mine	Nearly non-magnetic housing; timed fuse settings; once emplaced and activated cannot be removed.
	M70	Acoustic influence seabed mine	Employs highly sensitive sensors with a large explosive charge; can be laid from multiple platforms.
	M-71	Limpet mine	Attached magnetically and mechanically; fitted with a time fuse.

N/A = information not available or not provided.

* All known Russian mine types are not discussed here; for a more complete list refer to Jane's Underwater Warfare Systems.

Source: Watts, 1999

TABLE 2-2 Continued

Length (mm)	Width (mm)	Height (mm)	Weight (kg)	Charge (kg)	Operational Depth (m)
1,015	800	385	190	105	5-100
1,125	660	1,125	450	80	20-200
2,000	750	750	800	80	20-200
1,700	600	700	700	200-400	N/A
1,440-2,540	533	533	569	350	5-200
2,500	533	533	990	600	10-200
2,250-3,000	844-1,060	844-1,060	542-1,010	300-576	45-350
2,250-3,830 115A 450	400-630	400-630	240-926 115A 61	24-204	45-91
2,200-3,800	271-533	271-533	227-908	89-202	<100
3,700	533	533	1,075	45	>305
4,090	485	485	745	N/A	<100
670	430	320	50	27	30
2,823	534.4	534.4	1,000	700	50-150
345	345	245	14	3	30

MINE COUNTERMEASURES

Mines are difficult to detect in the complex nearshore environment, as they are relatively small, easily concealed, and can be laid from nearly any platform. This is particularly true for modern mines that have MCM-resistant features, such as anechoic coatings, non-ferrous components, and unconventional shapes. Often, complex and expensive equipment is needed to efficiently and effectively hunt and neutralize mines. As a result, countering mines requires financial expenditure out of proportion to the size and cost of a typical sea mine. This imbalance is increasing as mines become more sophisticated.

MCM can be both passive and active (Box 2-4). Passive MCM relies on limiting the influence of the ship's capability to be sensed by a mine. This is generally accomplished by reducing emitted acoustic and magnetic and electric signals from the vessel.

Active MCM includes both minesweeping and minehunting. Influence minesweeping uses acoustic, magnetic, and pressure signals to detonate emplaced mines. Mechanical sweeping uses a towed sled or other type of device to cut the cables of moored mines. These mines then float to the surface and are detonated by other means.

Minehunting is the safest and most effective method of dealing with mines, particularly with modern influence mines. Unlike minesweeping, where techniques need to be altered with variations in mine activation methods, minehunting is less dependent on the technical characteristics of the mine. Minehunting primarily relies on sophisticated high-frequency sonars, high-fidelity sidescan sonars, and magnetometers; either towed, hull-mounted, or hand-held. Once mines are located, they are destroyed using mine neutralization vehicles (Figure 2-3).

Oceanic conditions greatly influence offensive mining and MCM operations. Variations in environmental parameters, such as bathymetry, salinity, temperature, tidal range, currents, water clarity, and seafloor character, can alter and significantly degrade sensor performance and reduce operational capabilities. Many of the parameters listed above are collected in the battlefield by MCM forces. Detailed route surveying by the Naval Oceanographic Office (NAVOCEANO) outside periods of conflict provides data to support in situ battlefield measurements. Route surveys contribute a baseline image of passage routes, thereby reducing operational workloads to identification of changes in a baseline environmental description.

STRUCTURE OF MINE COUNTERMEASURE FORCES

Recently, the U.S. Navy reassessed its MCM strategy of maintaining only special purpose MCM forces and is now developing MCM capabilities that will be an integral part of carrier battle groups and amphibious ready groups (Box 2-6). These forward-deployed MCM capabilities have been defined as "organic" (Box 2-6). This change from an entirely dedicated MCM force to a combination organic and dedicated MCM force has occurred, in part, because of fiscal requirements to combine or develop multi-purpose assets and also because logistically dedicated MCM forces are often not stationed near areas of conflict and thus are unable to reach the battlefield in a timely fashion.

BOX 2-4 Mine Countermeasures

Passive

- Limit all influence signals emitted from target vessels.

Active

- Minehunting – The search for mines using magnetic, visual, and sonic means either through the use of divers, unmanned undersea vehicles (UUVs), or hull-mounted and towed instrumentation.
- Minesweeping – Detonation using both mechanical and influence methods. Minesweeping methods are highly dependent on mine characteristics and need to be continually altered for new mine types.

FIGURE 2-3 The Mine Neutralization Vehicle is used to destroy mines once they are located.

BOX 2-5 Goals for Naval Mine Countermeasures (MCM)

- Elevate status of MCM in naval operations:
  - Full commitment of naval leadership to well-equipped and trained MCM forces.
  - Treat MCM as an equal among major warfare mission areas.
  - Provide more significant MCM personnel training.
  - Integrate MCM, totally and realistically, into joint and fleet training exercises.
- Develop a Command and Control, Communications, Computers, and Intelligence (C^4I) architecture that supports the full range of MCM operations.
- Develop supporting and organic MCM systems that are capable of:
  - rapid deployment and employment,
  - high-area search rate with low false alarm generation,
  - rapid and wide-area detection, classification, and identification of mines,
  - automatically adapting to the environment,
  - autonomously destroying mines, and
  - supporting avoidance and in-stride mine and obstacle breaching from deep water to inland objectives.
- Develop an all-source precision database with the capability to provide real-time environmental assessments and forecasts and make it available to all MCM forces.
- Develop self-protection measures, including mine avoidance, signature manipulation, and shock hardening.

Source: Rhodes and Holder, 1998

BOX 2-6 Dedicated and Organic MCM forces

Dedicated – land, air, surface, or sub-surface forces trained and equipped specifically for and focused on MCM operations.

Organic – capability that is integral to forward-deployed forces to allow early MCM operations.

Source: Rhodes and Holder, 1998

Dedicated Mine Countermeasures

Dedicated MCM forces are the traditional special purpose units assembled for minehunting duties. Despite the Navy's recent moves to improve organic MCM capabilities, dedicated forces are the backbone of U.S. Naval MCM operations. These forces are highly trained with more advanced capabilities than exist in the organic MCM forces. Dedicated forces are able to reinforce organic MCM forces when the mission demands capabilities beyond those of organic systems.

Dedicated MCM forces rely on the combined assets of surface MCM ships, airborne MCM helicopters, and Navy Explosive Ordinance Disposal (EOD)/Naval Special Warfare (NSW) forces to ensure effective mine detection, classification, sweeping, and neutralization (Table 2-3; Box 2-7).

There are three classes of vessels in the dedicated MCM fleet. The largest of these is the USS *Inchon* (MCS-12; Table 2-3; Figure 2-4). The *Inchon*, originally commissioned in 1970, was redesigned in 1996 to provide support for sustained MCM operations by furnishing a landing platform for MH-53E Sea Dragon minesweeping helicopters and also to act as a repair and re-supply facility for Avenger and Osprey class coastal minehunters. The Avenger (MCM-1) class was designed to find, classify, and destroy moored and bottom mines in both coastal and offshore areas (Table 2-3; Figure 2-5). The ships use a fiberglass-sheathed wooden construction to make the hull less detectable by influence mines. Avenger class sweepers employ sonar and video systems for minehunting and use remotely controlled mine detonating devices, cable cutters, and other more conventional minesweeping measures for mine removal. The smaller Osprey class (MHC) has a shorter endurance and thus depends on support ships or shore-

TABLE 2-3 Primary Vessels of the Dedicated Mine Countermeasures Fleet

Craft	Crew (Officers/Enlisted)	Length (feet)	Beam (feet)	Displacement (full load; tons)	Draft (full load; tons)	Speed (knots)
USS *Inchon* (MCS-12)	45/634	602.3	112	18,340	26	23
Avenger class (MCM-1)	8/76	224	39	1,312	13	14
Osprey class (MHC-51)	5/46	188	36	893	13	10

*Note: When fully loaded USS INCHON can have in excess of 1300 personnel.
Source: Mine Warfare Command, 1999

BOX 2-7 Dedicated Mine Countermeasures (MCM) Mission

Provide the most maneuver area for joint and naval assets in the least amount of time with the least amount of residual threat to transitors while incurring minimal risk to MCM platforms.

FIGURE 2-4 The USS *Inchon* provides support for sustained mine countermeasure (MCM) operations by furnishing a landing platform for MH-53E Sea Dragon minesweeping helicopters while also acting as a repair and re-supply facility for Avenger and Osprey class coastal minehunters.

FIGURE 2-5 The Avenger class of minesweeper was designed to find, classify, and destroy moored and bottom mines in both coastal and offshore areas. This class has added capabilities to conduct mechanical, acoustic, and magnetic mine sweeping. The ships have a fiberglass-sheathed wooden construction to make the hull less detectable by influence mines.

FIGURE 2-6 The Osprey class has a shorter endurance than the Avenger and is more operationally capable due to its lower acoustic and magnetic signature. The Osprey only has a minehunting capability. Osprey class ships are constructed from glass-reinforced plastic (GRP) fiberglass.

based facilities for re-supply (Table 2-3; Figure 2-6). The small size of the Osprey MHC ships limits them to operations in shallow coastal waters and harbors. Osprey class ships are constructed from glass-reinforced plastic (GRP) fiberglass and have minehunting capabilities similar to the Avenger MCM ships. Unlike the MCM-1 class, the MHC is not equipped with minesweeping gear and is solely reliant on its minehunting capabilities.

In addition to the above vessels, there are other systems that are essential to the dedicated MCM force (Box 2-8). Dedicated airborne MCM (AMCM) forces rely on the MH-53E Sea Dragon class of helicopters (Figure 2-7)

FIGURE 2-7 The MH-53E Sea Dragon class of helicopters have minehunting and sweeping capabilities using a variety of towed gear.

BOX 2-8 Other Dedicated Mine Countermeasure (MCM) Assets

Airborne MCM (AMCM): MH-53E

- 24 HM-14/HM15 aircraft
- 450 NM range
- Five-seven person crew
- Mechanical sweep (Mk 103)
- Influence Sweep (MK 104, Mk 105)
- Minehunting (AQS-14 sonar)

Underwater MCM (UMCM): Explosive Ordinance Disposal (EOD)

- 17 mobile MCM detachments
- A detachment consists of one officer and seven enlisted personnel
- PQS-2A sonar
- Mk 16 non-magnetic underwater breathing apparatus

Underwater MCM (UMCM): Marine Mammal System (MMS)

- 12 mammals
- Four systems available and one under development
  (1) *Mk 4* – employs four dolphins for close-tethered, deep-moored minehunting and neutralization. The Navy employs this capability to neutralize all buoyant mines.
  (2) *Mk 5* – employs four sea lions for acoustic object recovery. Sea lions attach recovery pendants to exercise mines, torpedoes, and other objects.
  (3) *Mk 6* – employs six dolphins for swimmer and diver detection and defense of harbors, anchorages, and individual ships.
  (4) *Mk 7* – employs eight dolphins for mine detection, classification, location, and neutralization of bottom mines.
  (5) *Ex 8* (under development) – will employ six dolphins for exploration and reconnaissance of in-volume moored and bottom mine-like contacts in the very shallow water (VSW) zone.

Source: Mine Warfare Command, 1999

whose minehunting and sweeping capabilities depend on towing a variety of gear, such as minehunting systems and minesweeping (mechanical/influence) equipment. Sea Dragon helicopters have a four-hour mission capability.

Underwater mine countermeasures (UMCM) depend on both the EOD units and specially trained Marine Mammal System (MMS) mammals, which perform tasks similar to those of the EOD units with high confidence and greater speed. The MMS force is the only asset in the fleet that can detect buried mines. EOD and MMS systems can together accomplish a range of tasks from mine search through identification. In the future EOD and MMS teams will be replaced by Unmanned Undersea Vehicles (UUV), which are not limited by endurance, particularly in cold or otherwise hostile environments. UUVs also have the advantage of removing the warfighter from the battlefield.

Organic Mine Countermeasures

An important component of a successful MCM program is the capability of carrier battle groups, amphibious ready groups, and other naval assets to deal with naval mines while dedicated MCM forces are in transit or otherwise occupied. Without this capability, forward-deployed forces must remain idle while waiting for mine-clearing assistance. To ensure that commanders of forward-deployed assets are able to operate freely in the nearshore battlefield, the Navy has developed organic MCM strategies (Box 2-9). The goal of organic MCM is to enable Naval forces to assess the danger posed by a potential minefield, reduce this risk to an acceptable level, and concentrate on their primary mission. By 2005, the first aircraft carrier battle group will no longer rely solely on dedicated minesweep-

BOX 2-9 Organic Mine Countermeasure (MCM) Strategies

Detect presence or absence of mines. Perform basic exploratory and reconnaissance operations
- Has mining occurred?
- To what extent and density?
- Where are the boundaries and gaps?

Operational maneuver options
- Verify and avoid mined operating areas (OPAREA)
- Exploit gaps
- Limited clearance and neutralization (punch through)

Organic limitations include:
- MDA follow-on clearance
- Mine exploitation
- Sea lanes of communication (SLOC)/OPAREA expansion
- Amphibious assault support

ers. Instead, each battle group and amphibious ready group will have air, sea, and undersea capabilities to hunt and destroy mines.

In the future, naval organic MCM capabilities include not only on- and off-board sensors but also the operational infrastructure necessary to plan MCM operations and implement tactical decisions to reduce the risk of operating in mined waters. Tools used by organic MCM forces include H-60 helicopter-equipped AMCM systems, surface-ship-deployed mine reconnaissance capabilities, and submarine-deployed systems. There is no single piece of equipment that can provide all operational capabilities needed; thus organic MCM forces are using and developing three underwater reconnaissance systems, two airborne mine detection systems, and two airborne mine destruction systems. Each of these systems will have the capability of dealing with a small number of mines; operated together these systems will be able to cope with the operational challenges posed by the full range of naval mines.

A significant organic MCM development is the ship-deployed Remote Minehunting System (RMS) that was developed to provide battle groups and surface combatants with a means of detecting and avoiding mines from deep to very shallow water. This remotely operated system will use computer-aided detection and precise navigation systems for mine location and classification. Other organic MCM systems in development include the Airborne Mine Neutralization System (AMNS), which is a remotely operated, expendable, torpedo-like device capable of identifying and neutralizing bottom and close-tethered mines; the Airborne Laser Mine Detection System (ALMDS), an electro-optical mine detection system that uses an aircraft-mounted laser to detect floating and near-surface mines; and the Rapid Airborne Mine Clearance System (RAMICS), which provides near-surface mine neutralization using a gattling gun system.

Battlefield reconnaissance data are essential for safe nearshore operations. Thus, plans are in place to develop and expand organic MCM capabilities with both Near-term and Long-term Mine Reconnaissance Systems (NMRS and LMRS). The NMRS will be operational first and will provide limited mine detection, classification, and localization from a UUV launched and recovered from the torpedo tube of a Los Angeles class submarine.

MINE WARFARE DOCTRINE

Naval Warfare Publication 3-15

Naval Warfare Publication (Department of the Navy, NWP 3-15.41) "Mine Countermeasures (MCM) Planning and Procedures" provides operational planning and calculation procedures for use in the formulation and evaluation of MCM operations. Under current procedures, environmental data, along with available intelligence data, are merged with system performance data to plan an MCM mission.

The full spectrum of environmental parameters affecting sensor and system performance are not consistently monitored during MCM operations. Of the environmental data collected, only those describing seafloor properties are considered within NWP 3-15 as these parameters strongly impact the level of backscattered acoustic energy and thus, significantly affect sonar performance. To support minehunting sonars, bottom composition, bottom roughness, percent impact burial, and NOMBO densities are extracted or estimated from available meteorological and oceanographic data holdings. In accordance with NWP 3-15, sediment grain size is limited to three groupings: mud, sand, or rock, and bottom roughness is characterized as smooth, moderate, or rough. With the high-sonic frequencies used by minehunting sonars, bottom features (i.e., ridges, sand waves, etc.) in excess of 0.5 m are defined as rough. NOMBO density is the estimate of mine-like bottom objects occurring in a square nautical mile (nm^2). NOMBO densities of 0-14 $/nm^2$ are categorized as clutter density 1; densities from 15 to 40 NOMBOS/nm^2 are categorized as clutter density 2; and densities in excess of 40 NOMBOS/nm^2 are classified as clutter density 3. NOMBO density influences the amount of incorrectly identified mines and the time needed to complete a mission. In remote regions, bottom roughness and NOMBO density are often difficult to assess, as the definition of these parameters requires high-resolution bathymetric and sidescan sonar surveys, which are not always available.

Once bottom composition, estimation of mine burial, bottom roughness, and NOMBO density data are acquired, they are combined to form a mine warfare bottom category in accordance with NWP 3-15. Bottom categories range from a best case "A" defined as an optimum minehunting seafloor, to the worst case "D," typified by a potential for high-mine burial. When combined with NOMBO density estimates, the seafloor can be categorized

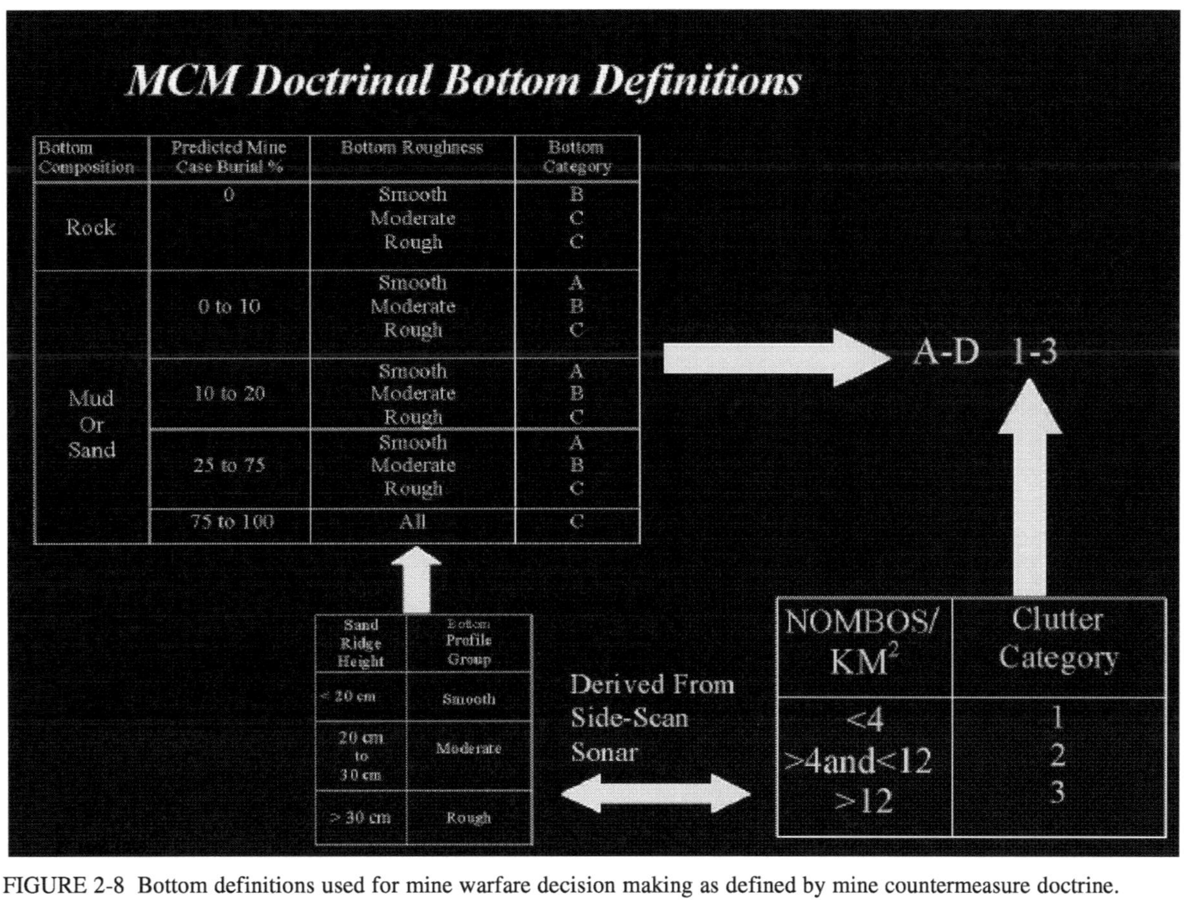

FIGURE 2-8 Bottom definitions used for mine warfare decision making as defined by mine countermeasure doctrine.

from "A1" to "D3." This estimate is used to obtain a value for the sonar detection width "A" and probability of mine detection "B" referenced from NWP 3-15. After "A" and "B" values are determined, the MCM operator decides the number of ship tracks (or lanes) and track spacing necessary to achieve an acceptable safe clearance level (Figure 2-8).

Shortfalls in NWP 3-15

Although it is well known that the environment plays a pivotal role in MCM operations, current doctrine (NWP 3-15) does not adequately account for the highly variable conditions often encountered in MCM operational areas. NAVOCEANO maintains high-resolution geologic, bathymetric, and oceanographic databases in support of fleet mine warfare efforts, yet the full range and resolution of these data are not typically exploited. For example, sonar width "A" is highly dependent on potentially variable seawater physical properties. Despite this, sonar width is a fixed quantity in NWP 3-15 for all environments. Consequently, environmental variations can result in data voids, termed "holidays," because the assumed sonar detection width "A" is greater than the actual sonar detection width. Conversely, redundant coverage (wasted time) will result when the assumed width "A" is less than the sonar detection width for the environment in question.

FUTURE TRENDS IN MINE WARFARE

To ensure that naval forces are able to maneuver and control operations in the nearshore environment, there should be continued development of mine laying and mine countermeasures, while also incorporating advancements in environmental predictive capabilities to maximize the effectiveness of mine warfare operations. This can occur only with increased integration of mine warfare into fleet training, exercises, and deployments. Despite the strategic benefits of mine warfare, mine warfare has undergone periods of relative neglect between conflicts. Currently, budget increases are enabling the Navy to overhaul its mine and MCM operations. Future funding increases are also a possibility, as the President has requested $64 million for research and development of MCM (Abel, 1999).

The numerous planned initiatives to refine the capabilities of mine warfare (MIW) forces involve improved hardware and support for the MIW community. With respect to hardware, one future direction is the refurbishment of outdated mines to extend their operational life, using upgrade kits that modernize the firing mechanism while retaining the original mine case and warhead.

Numerous new technologies are also emerging that will improve the effective range, countermeasure resistance, reliability, and versatility of naval mines. A significant advancement is the propelled warhead (PW) mine that uses buoyancy or propulsion systems to transport a warhead to its target. PW mines have many benefits over more conventional explode-in-place (EIP) mines (see "Mine Classification" section, earlier in this chapter): greater range, increased speed to target (reduces the time available for the target to initiate countermeasures or evasive maneuvers), reduced need for a large number of mines, improved target detection, and deep water operational capabilities where mechanical minesweeping is difficult.

New countermeasure-resistant mine designs include fiberglass and non-ferrous cases that reduce maintenance and detection. These cases can also have anechoic coatings to reduce their sonic signature. New mines are also being designed with unconventional shapes, making them difficult to identify on seafloor images. Mines are being fitted with high-sensitivity multiple-influence detonation systems with delayed sensors to fire a preset time after a triggering signal. A significant advance is the self-burying mine. This mine will dramatically reduce the effectiveness of current minehunting sonars and cause an increased dependence on slower minesweeping techniques. Future advancements in remote control (RECO) systems will enable mines to be turned on and off when needed and enable them to be detonated remotely. RECO-equipped mines will provide an important tactical advantage, particularly in shallow water and in shipping channels, where it may be necessary to deactivate the mines to allow certain ships to pass.

Advances in minehunting and sweeping are also underway. UUVs will increase the capability to search an area for mines without compromising intent and position. These vehicles will incorporate a variety of sensors to survey nearshore operational areas clandestinely and, if necessary, neutralize emplaced mines. This advance will remove personnel from the dangerous job of minehunting and destruction. However, to increase the effectiveness of UUVs, greater speed and range and deeper diving capability is needed (Morison, 1995). In water depths greater than 10 ft., the diesel-powered RMS discussed previously will eventually provide the dual capacity to locate and neutralize mines. The next RMS prototype will have an over-the-horizon communications link that will give operational commanders the ability to remotely control the RMS at a range of up to 100 miles. Systems to be used with the multipurpose H-60 helicopters are also under development. These advances in minehunting and sweeping capabilities will greatly reduce the time needed to classify and neutralize mines in the coastal zone.

New systems are also being developed for use in the surf zone. The Shallow Water Assault Breaching System (SWABS) and Distributed Explosive Technology (DET) can deliver an array of explosives to remove mines and obstacles ahead of a landing force. These systems have the advantage of an increased standoff distance, reducing the danger to platforms and personnel.

Advances are also being made in the area of sensor development. For example, the Tactical Acoustics Measurement and Decision Aid (TAMDA) will not only give temperature and depth measurements, as with conventional XBTs, but will also collect salinity and ambient noise data for use in acoustic sensor calibration. TAMDA will also incorporate a broadband probe pulse to determine reverberation and bottom composition, will have a global positioning system (GPS), and will have a longer operational life to provide extended data collection capabilities.

In addition to hardware improvements, improvements to support systems for the MIW community are also planned. Many of these systems will increasingly rely on computer models and advanced oceanographic databases to store MCM-specific types of digital information. An important development will be a shared database containing oceanographic data from all available sources. These data will be used to update and refine environmental assessments and will be the basis of a distributed database digitally available to the end user. Furthermore, recent developments in advanced visualization hardware and software systems will enable the mine warfighter to interact, in real time, with 3D graphical representations of oceanographic data. This capability will enhance the warfighter's ability to understand local environmental variability and thus will greatly benefit mine warfare operations.

The MIW community will also benefit from advances in data processing algorithms. For example, continued development of mathematical inversion techniques for use on sonic datasets will provide qualitative estimates of bottom type and composition from sonar data. In addition, advances in data processing from sensors, such as LIDAR, will provide high-resolution seafloor bathymetry data and enhance the ability to identify mine-like features in seafloor images.

3

Oceanography and Mine Warfare

Numerous meteorological and oceanographic parameters are important for mine warfare operations (Box 3-1). At present, some of these are part of routine data collection for the Naval Oceanographic Office (NAVOCEANO) and mine countermeasure (MCM) forces (Boxes 3-2 and 3-3). The importance of each of these oceanographic parameters for mine warfare operations vary, in some cases significantly, between different water depths (Table 3-1). For example, in many coastal environments, wind-generated currents are of little importance in the surf zone as they are generally overwhelmed by wave-induced currents, whereas in very shallow water (VSW) and shallow water (SW) zones information on wind-generated currents is essential.

OCEANOGRAPHY OF THE NEARSHORE ENVIRONMENT: THE CHALLENGE FOR MINE WARFARE

A significant challenge for mine warfare operations in the nearshore environment is the shortened spatial and temporal scales of oceanic and atmospheric variability. This complexity makes it difficult to predict environmental changes and their resultant effect on sensors, platforms and weapons (Rhodes and Holder, 1998). Oceanic variability in the coastal zone can result from ocean fronts, freshwater inputs (precipitation and river runoff), local and

BOX 3-1 Important Environmental Characteristics That Influence Mine Warfare Operations

- Meteorology
  - Wind
  - Air temperature/humidity
  - Cloud cover/precipitation
- Oceanography
  - Bathymetry
  - Temperature/salinity profile
  - Water clarity
  - Tides
  - Currents
  - Acoustic background
  - Sea state
  - Upwelling
  - Biologics
  - Sediment/bottom type

Adapted from Rhodes and Holder, 1998

BOX 3-2 Commander, Mine Warfare Command (COMINEWARCOM)–Acquired Environmental Data

- AN/SQH-4 Battlespace profiler data – Conductivity, temperature, and depth (CTD) measurements; interfaces with expendable bathythermographs (XBT) and expendable sound velocity (XSV) probes.
- AN/AQS-14 Acoustic imagery – Helicopter deployed multi-beam minehunting sidescan sonar.
- AN/SQQ-32 Minehunting sonar – Variable depth sonar installed on mine countermeasures (MCM) and mine-hunter, coastal (MHC) class ships.
- Explosive Ordinance Disposal (EOD) Diver observations – Qualitative environmental descriptions, including arm thrust measurements of the seafloor, bottom samples, horizontal visibility measurements, and estimates of bottom currents.

BOX 3-3 The Naval Oceanographic Office (NAVOCEANO) –Acquired Environmental Data

- Acoustic imagery – Processing of AN/AQS-14 imagery.
- Clutter density – Determined as part of AN/AQS-14 image processing.
- Currents and provinced currents – Data sources are from moored current meters, acoustic doppler current profilers (ADCP), and diver observations.
- Soundings and gridded bathymetry – Single-beam and multi-beam data used to generate sounding files or gridded bathymetry.
- Optics – Measured parameters include optical transmission loss, absorption, and irradiance.
- Biologics – Bioluminescence, dangerous marine animals, and biofouling.

TABLE 3-1 The Most Common Oceanographic Datasets Used for Nearshore Mine Warfare Operations and Their Importance in the Different Water Depth Zones Defined by the Navy

	Surf Zone	VSW	SW	Deep Water
SEAFLOOR				
Bathymetry	H	H	H	M
Sediment grain size	M	H	H	M
Seafloor clutter density	M	H	H	L
Bottom roughness	M	H	H	L
Mine burial	H	H	H	L
WATER COLUMN				
Currents	L/H*	H	H	M
Water clarity	L	H	H	M
Temperature and salinity	L	M	H	H
Waves	H	H	M	L
Acoustic properties	N/A	H	H	H

*Within the surf zone, information on wind-generated currents is of low priority as these currents are generally overwhelmed by wave-generated currents.
VSW = Very Shallow Water
SW = Shallow Water
H = High, a parameter that is essential for mine warfare operations in this depth zone;
M = Medium, a parameter that is useful for mine warfare operations in this depth zone;
L = Low, a parameter that is of little use for mine warfare operations in this depth zone;
N/A = A parameter that is not applicable or not measurable in this depth zone.

synoptic weather disturbances, tidal fluctuations, sea and land ice, and internal and coastal trapped waves. In the open ocean, fronts vary on time scales of days to weeks, whereas in the coastal ocean they can vary hourly as a result of tidal forcing and local weather phenomena (Martin, 1997). These high-frequency variations have significant impacts on many aspects of mine warfare (MIW) operations, the most important being sensor performance, mine burial capability, and the mobility of MCM forces in shallow water.

Oceanic complexity increases in the surf zone due to the dynamic interaction of the water column with the seafloor and the adjacent landmass. This interaction modifies waves, localizes alongshore currents, influences underwater structures, and forces resuspension of sediments. These factors make data collection and remote-sensing in the surf zone environment extremely difficult.

To ensure continued U.S. Naval domination in littoral mine warfare, it is critical that naval operators have access to tactical decisions aids (TDA) based on sophisticated environmental datasets and results from quantitative models of nearshore oceanographic processes. Numerical models traditionally used in the open ocean are of little use in the coastal zone because of the high-frequency variability in this environment. Current shallow water and surf zone modeling capabilities are not well developed. To advance these models, it is imperative that progress be made in understanding coastal environmental processes.

The ability to process, interpret, and use environmental data for mine warfare operations is critically dependent on high-quality advanced training for mine warfare forces. This need will continue to increase with greater dependence on complex environmental datasets and predictive models. It is thus imperative that concurrent advances occur in warfighter technical training and environmental data collection.

THE ROLE OF OCEANOGRAPHY IN MISSION PLANNING, BATTLEFIELD PREPARATION, AND MISSION SUCCESS

The ocean environment significantly influences mine warfare tactical planning. An in-depth understanding of nearshore environmental variability is important, not only for physical mine placement but also for its impact on minehunting sensors. Despite the complexity of the coastal environment, Naval Meteorological and Oceanographic (METOC) infrastructure is adapting to the challenge of operations in these water depths. Specific goals being addressed are (1) remedying the lack of high-resolution nearshore oceanographic and bathymetric data, particularly in areas of potential conflicts and (2) developing better in situ sampling systems for measurement of tactically significant oceanographic and meteorological parameters.

As discussed previously, the need to improve environmental databases and interactive data visualization capabilities is particularly important for mine warfare planning and operations. The massive databases collected in open ocean environments during the Cold War are an important start, but dedicated nearshore survey data are an essential supplement to existing data. The Director, Expeditionary Warfare (OPNAV N85) and Mine Warfare Command (COMINEWARCOM) are providing resources for developing these databases at a sufficient resolution for MCM applications. This work will require extensive high-resolution, peacetime oceanographic surveys to be completed in high-priority areas (Martin, 1997). Tactical decision aids (discussed below) developed from nearshore oceanographic databases, when modified by in situ data, will provide a powerful tool for the mine warfighter to optimize operational capabilities.

Mission Planning

In current mine warfare mission planning, historical oceanographic data are acquired from the NAVOCEANO database. These data, including seafloor characteristics, water column properties, and atmospheric parameters, are input into the Mine Warfare Environmental Decision Aid Library (MEDAL) to determine optimal lane spacings and predict operational time lines and risk. Once into the exercise, in situ data are collected and input into MEDAL and other TDAs. When in situ data are entered into MEDAL, the operator can output predictions of best- and worst-case scenarios for line spacings to balance operational objectives with clearance time and risk. It is important at this stage of operations that mine warfare personnel be receptive to modifying tactics to fit changes in environmental conditions.

In the future, there will be increasing reliance on battlefield sensors for MCM operations. The effectiveness of these sensors is highly dependent on environmental conditions. Understanding the environment in which the sensor is operating and the effects of environmental variability on measurement parameters ensures maximum sensor performance and utility of the resulting data. Because of this requirement, OPNAV N85 is installing the Battlespace Profiler (BSP), a tool similar to a Conductivity, Temperature, Depth (CTD) profiler on all MCM vessels to enable them to measure temperature, salinity, and depth. In part, this requirement will increasingly be met by modern sensors, which maximize system performance by adjusting to environmental conditions.

Advances in sensor and communications technology have provided an important support tool for the warfighter, enabling rapid environmental assessments and vastly improving how naval forces collect and distribute battlefield information. Advances in electronic processing speed and miniaturization have enabled the development of new sensors for Unmanned Undersea Vehicles (UUVs), while innovative image processing of electromagnetic and sonar data is enabling enhanced visualization of the battlefield. In addition, recent advances in both observational sampling systems and numerical modeling techniques make the large amounts of environmental data being collected more accessible to the MCM community. An improved understanding of the circulation, seawater properties, and bottom characteristics of nearshore areas depends upon the success with which these observed and simulated data sets are managed and integrated with visualization and analysis technologies.

SUPPORT FOR MINE WARFARE: INFORMATION AND SYSTEMS

Navy METOC is moving to meet the challenge of nearshore operations by investing resources in new technologies and techniques for characterization of oceanographic variability in the coastal zone. An in-depth understanding of the coastal zone environment will help ensure that naval forces are able to operate freely in these areas.

As part of the Mine Warfare Campaign Plan, COMINEWARCOM and NAVOCEANO have coordinated their efforts to provide an enhanced capability for environmental analysis. As an example, a Route Survey Environmental Data Base Concept of Operations (CONOPS) has been developed to outline how this will be accomplished and how it will provide enhanced environmental capabilities to warfighters at COMINEWARCOM.

Data types presently used include high-resolution sidescan sonar acoustic images collected with Airborne Mine Countermeasures (AMCM) assets; Explosive Ordinance Disposal (EOD) diver reports; and salinity, temperature, and sonic velocity data collected by the BSP. After collection, these data are forwarded to NAVOCEANO, where they are processed and entered into the master data warehouse. NAVOCEANO analysts combine the field-collected data with historical data to develop a description of the environment in question. This analysis is then returned to COMINEWARCOM as a digital file that can be entered into MEDAL, the Comprehensive Environmental Assessment System (CEAS), the Naval Interactive Data Analysis System (NIDAS), and the Unified Sonar Image Processing System (UNISIPS). Data can also be returned to COMINEWARCOM as information contained in other NAVOCEANO products, such as Special Tactical Oceanographic Information Charts (STOICs), acoustic imagery mosaics, and Mine Warfare Pilots (MWP; see below). The aim of these systems and products is to help the warfighter determine the best way to employ sensors and weapons. Figure 3-1 summarizes the environmental data flow between COMINEWARCOM and NAVOCEANO.

Environmental Support Systems

Oceanographic Support Subsystem Data Warehouse

The Oceanographic Support Subsystem Data Warehouse (OSS DW) is both a tool for data location and a facility for management of data, information and software held at NAVOCEANO. A catalog function enables users to locate common software applications for execution or software release. The OSS DW is maintained through a graphical user interface that allows users to query geo-referenced information for a particular area of interest, using Structured Query Language (SQL), the graphical user interface, or a Web browser.

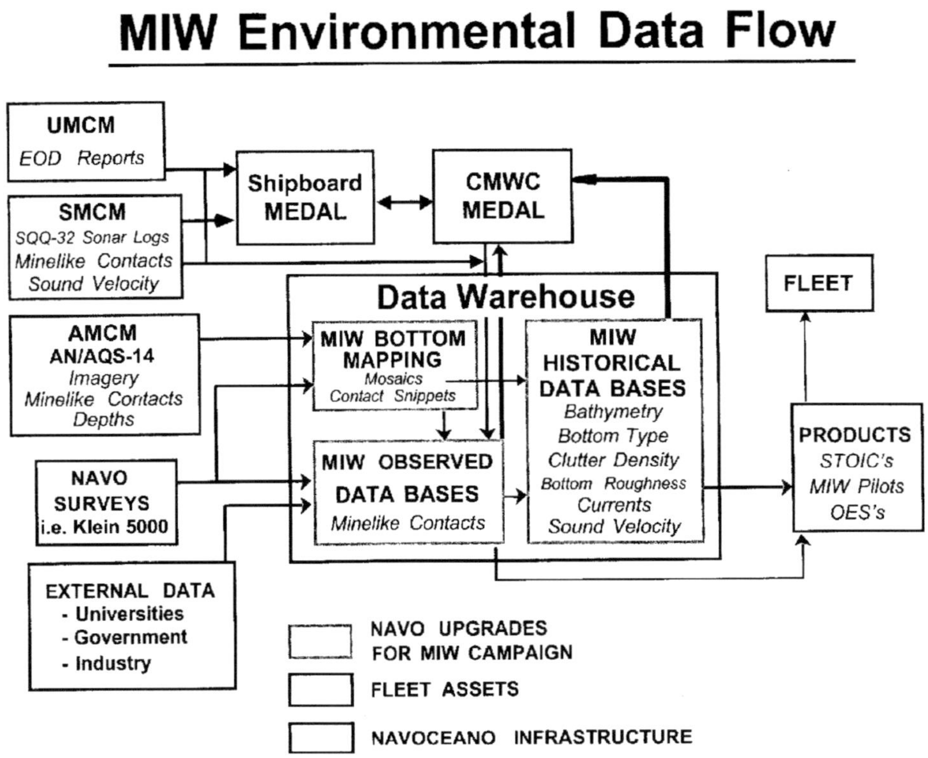

FIGURE 3-1 Mine warfare environmental data flow between Naval Oceanographic Office (NAVOCEANO) and Commander, Mine Warfare Command (COMINEWARCOM).

Mine Warfare Environmental Decision Aid Library

MEDAL uses historical and in situ environmental data for mine warfare mission planning, system performance evaluation, tactics development, and exercise evaluation. MEDAL integrates command and control capabilities, performance assessment models, algorithms, and environmental databases to determine mine threats, develop hunting and sweeping plans, determine an optimum force mix, and recommend tactics and techniques. In addition to planning, MEDAL is also capable of monitoring operations in progress, assembling in situ measurements for transfer to NAVOCEANO, and executing post-exercise reconstruction and analysis. MEDAL provides the MCM commander with the ability to manipulate the following datasets:

Bathymetry – Bathymetry is stored in MEDAL as individual points from EOD diver observations, the AQS-14 altimeter, soundings, and gridded bathymetric files.

Bottom Characteristics – At present, MEDAL uses 16 categories to describe sediment characteristics and there are plans to increase this to 72 in future builds (Figure 3-2). This information is acquired from historical databases resident at NAVOCEANO and can be augmented by in situ measurements from diver reports or indirectly from video on the AN/SLQ-48 Mine Neutralization Vehicle. Despite the fact that NWP 3-15 limits grainsize to three groupings, sediment grainsize is usually input into MEDAL using the four sediment types in acoustic performance models.

Currents – Surface and seafloor current speed and direction estimated from diver reports are an important component of operational planning. These data are entered into MEDAL as point, processed, or textual information. At present, there are no circulation models incorporated into MEDAL.

Seawater Physical Properties – Seawater temperature and salinity can be obtained from diver reports or CTDs deployed from MCM platforms. Seawater physical properties are mainly used for minehunting sonar performance

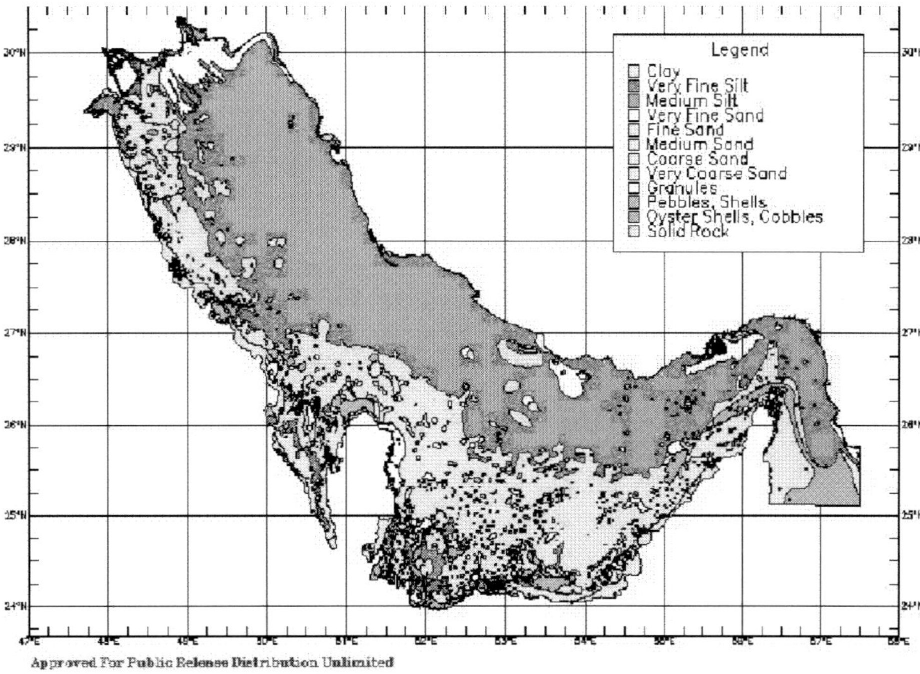

FIGURE 3-2 A typical example output of seafloor bottom types from the Arabian Gulf. Although it is difficult to distinguish between sediment grainsize shadings in the legend on the right, this figure shows the level of detail available for certain field areas.

models. In the future, MEDAL will also use these data to estimate the performance of other acoustic equipment, including the coverage of acoustic minesweeping gear.

Atmospheric Data – Wind speed and direction and ambient air temperature are the main atmospheric datasets collected. Wind speed is an important input to sonar performance algorithms used by AMCM operations and also for helicopter flight parameters, such as blade and pylon unfolding, rotor engagement, and launch. Wind speed is also a factor in sea state, and affects both Surface Mine Countermeasure (SMCM) and Undersea Mine Countermeasure (UMCM) operations. Future versions of MEDAL will be able to incorporate regional climatic databases.

Magnetic Capability – These data include measurements of seawater and sediment conductivity, electrical depth, and reflection coefficients used to calculate the performance of minesweeping systems.

Acoustic Data – Acoustic data entered into MEDAL are the K1 and K2 parameters (slope of a transmission loss curve up to and beyond 100 yards) used to estimate coverage for acoustic minesweeping equipment. At present, no algorithms exist to recalculate these parameters according to environmental changes, but future algorithms may take into account acoustic water depth, seawater physical properties, wind speed, and bottom composition.

Comprehensive Environmental Assessment System

CEAS provides multi-disciplinary environmental data assessment and analysis, survey and special deployments planning, and MIW support. CEAS was designed to provide a broad range of capabilities while also having the flexibility for future enhancements. Some of these capabilities are the production of acoustic image mosaics, data queries, the creation of derived data, and output of diverse digital and hardcopy export formats.

At present MEDAL is limited in its ability to input, manipulate, and display oceanographic data. Thus, the graphic capabilities of CEAS are being used at COMINEWARCOM and at sea on the USS *Inchon* as a briefing tool to subset data from large regional datasets and to view and manipulate sidescan sonar imagery.

Naval Interactive Data Analysis System

NIDAS is a Windows-based application that processes and analyzes temperature and salinity data from vertical profiles (historical or real time), gridded profiles (ocean models), or satellite sea surface imagery. NIDAS is designed to be used with shallow water, high-variability datasets typical of the coastal zone where statistics and modeling may be of little use. Specific uses for NIDAS include the development of survey sampling strategies, quality control of ocean profile data, construction of ocean climatological databases, oceanographic product generation (see below), and quality control of ocean thermal models. NIDAS supplements MEDAL, enabling the COMINEWARCOM METOC officer to construct synthetic profiles; perform three-dimensional grid slicing as waterfall plots or horizontal and vertical contours; export observed, provinced, or gridded profiles for use in sonar performance models; and input in situ CTD and Expendable Bathythermograph (XBT) data for comparison to historical master oceanographic observation datasets (MOODS) or gridded provinced databases.

Unified Sonar Image Processing System

UNISIPS was designed to standardize the processing of acoustic imagery data. UNISIPS can perform signal and image processing of raw data to create digital mosaics. UNISIPS can also use acoustic data to determine seafloor characteristics (e.g., clutter density and roughness), which are then entered into the bottom characteristics database. UNISIPS is being used as a real-time bottom-mapping analysis tool during exercises. This provides the METOC officer with extended capabilities, such as comparison of collected imagery to historical imagery to locate mine-like contacts.

Environmental Data Products

Acoustic Imagery Mosaics

Acoustic imagery mosaics are constructed in UNISIPS, and can be produced in either hardcopy or digital formats (Figure 3-3). Mosaics can then be input into CEAS, where additional environmental or operational information can be added. If a historical database of images is available, mosaics are an extremely useful tool for indicating changes in seafloor clutter which could indicate the presence of emplaced mines (Figure 3-4). Since CEAS is a geographic information system (GIS) application, layers of information can be compiled allowing the MCM commander to use the environment for tactical advantage. Acoustic imagery mosaics provide accurate estimates of bottom roughness and clutter density (Figure 3-5), and are source data for determining MIW doctrinal seafloor classifications. More important, imagery mosaics allow the MCM commander to refine operational areas based on observed topography. Since the ultimate goal is to reduce time lines and risk, operational areas and routes can be directed over more benign geologic areas where time lines and risk are lower.

Special Tactical Oceanographic Information Charts

STOICS are 1:25,000-scale bathymetric plots of a particular area of interest that are accompanied by environmental data collected over a specific time frame (e.g., month or season). This product has been designed to be a simple chart with the exact data needed by the user for nearshore mission planning. STOICS are not only used by the mine warfare community but also the special warfare and amphibious warfare communities.

Mine Warfare Pilots

MWPs provide environmental information essential for minefield and MCM planning and complement mine warfare doctrine. Specifically, MWPs include information on human activities, climatology, tides, currents, waves, sea ice, physical properties of seawater, seabed characteristics, and acoustic, magnetic, and biologic characteristics of a region of interest. As with STOICS and OESs (described below), MWPs are an essential planning tool for the COMINEWARCOM METOC officer. MWPs are used to define the MCM operational environment, including po-

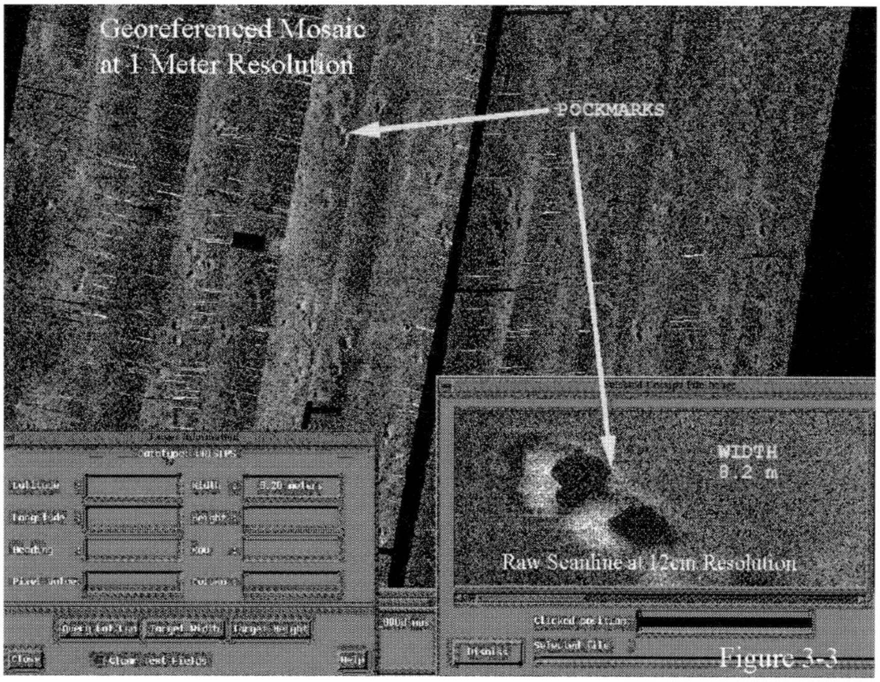

FIGURE 3-3 An acoustic imagery mosaic with one meter resolution. On this image numerous "pockmarks" can be seen. These features are likely to have resulted from seafloor release of hydrocarbon gases.

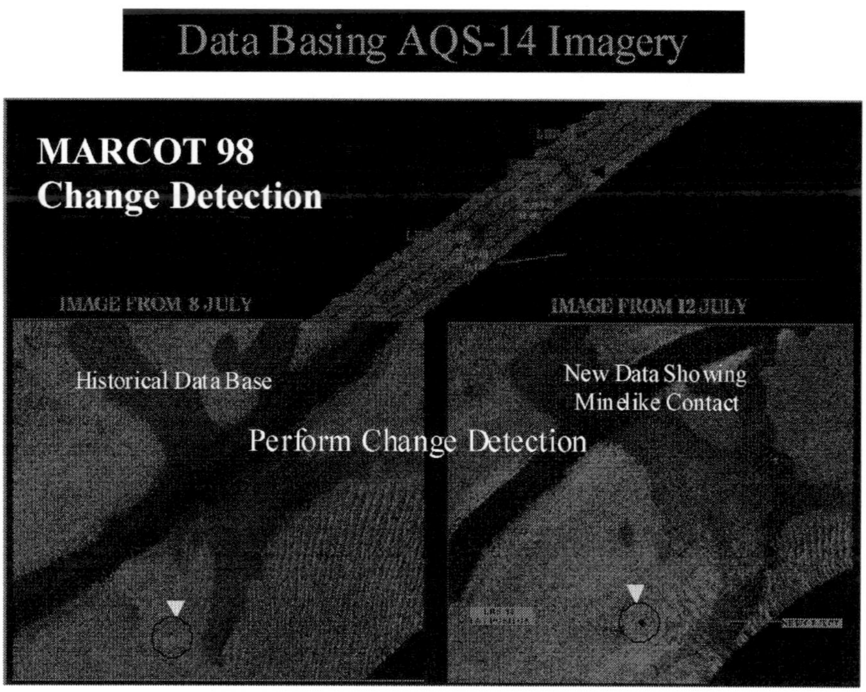

FIGURE 3-4 A pair of acoustic imagery mosaics collected four days apart. Comparison of the older image on the left and the newer image on the right shows the appearance of a mine-like contact in the time between the collection of the two images.

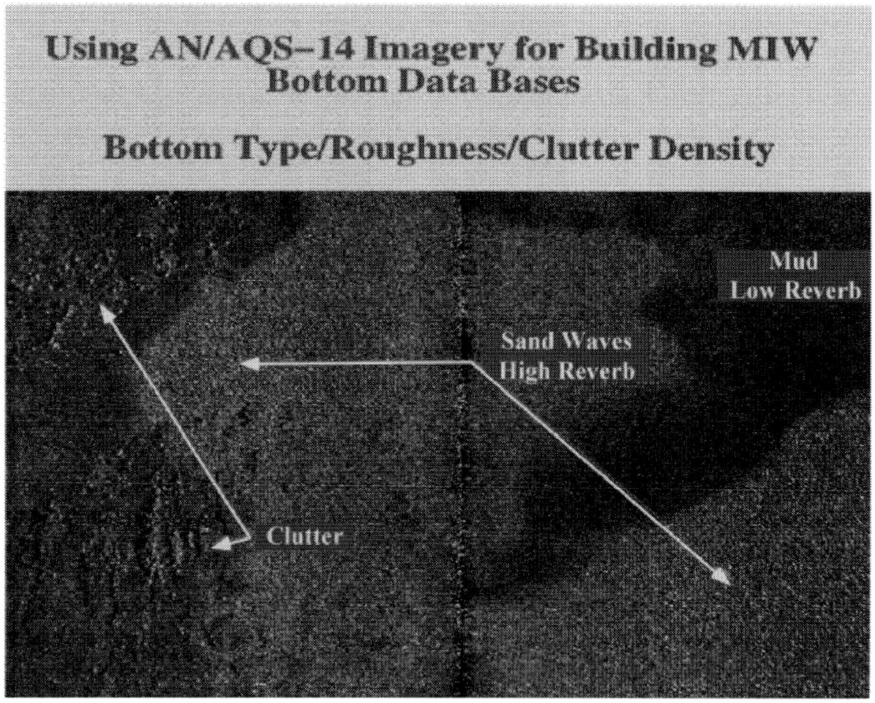

FIGURE 3-5 Acoustic imagery mosaics can be used to determine bottom type, roughness, and clutter density. In this image, an area of low reverberation, which has been interpreted as muddy sediment, can be seen in the upper right and high reverberation in the center of the image has been interpreted to reflect sandy sediment. An area of high clutter can be seen on the left.

tential limitations on equipment and platforms, and to supplement information and analyses generated by MEDAL, CEAS, NIDAS, and UNISIPS. MWPs are digitally output to the METOC officer.

Oceanographic/Environmental Summary

OESs provide baseline oceanographic information on specific geographic areas. These summaries can include physical oceanographic (currents, waves, tides, etc.), bathymetric, geomorphologic, biologic, and acoustic data. Many of the data types included on OES plots are similar to those on MWPs, but the latter generally cover a broader geographic region. As with MWPs, OESs are produced in digital format.

Tailored Requests for Planning and Exercises

NAVOCEANO also receives requests for a wide range of specific datasets to support MIW exercises. This tailored information provides an environmental description of the operational area as support for planning and briefings.

SUPPORT FOR MINE WARFARE: ENVIRONMENTAL SCIENCE PROGRAMS IN THE OFFICE OF NAVAL RESEARCH

The role of the Office of Naval Research (ONR) is to coordinate, execute, and promote science and technology programs in the U.S. Navy and Marine Corps through universities, government laboratories, and other research

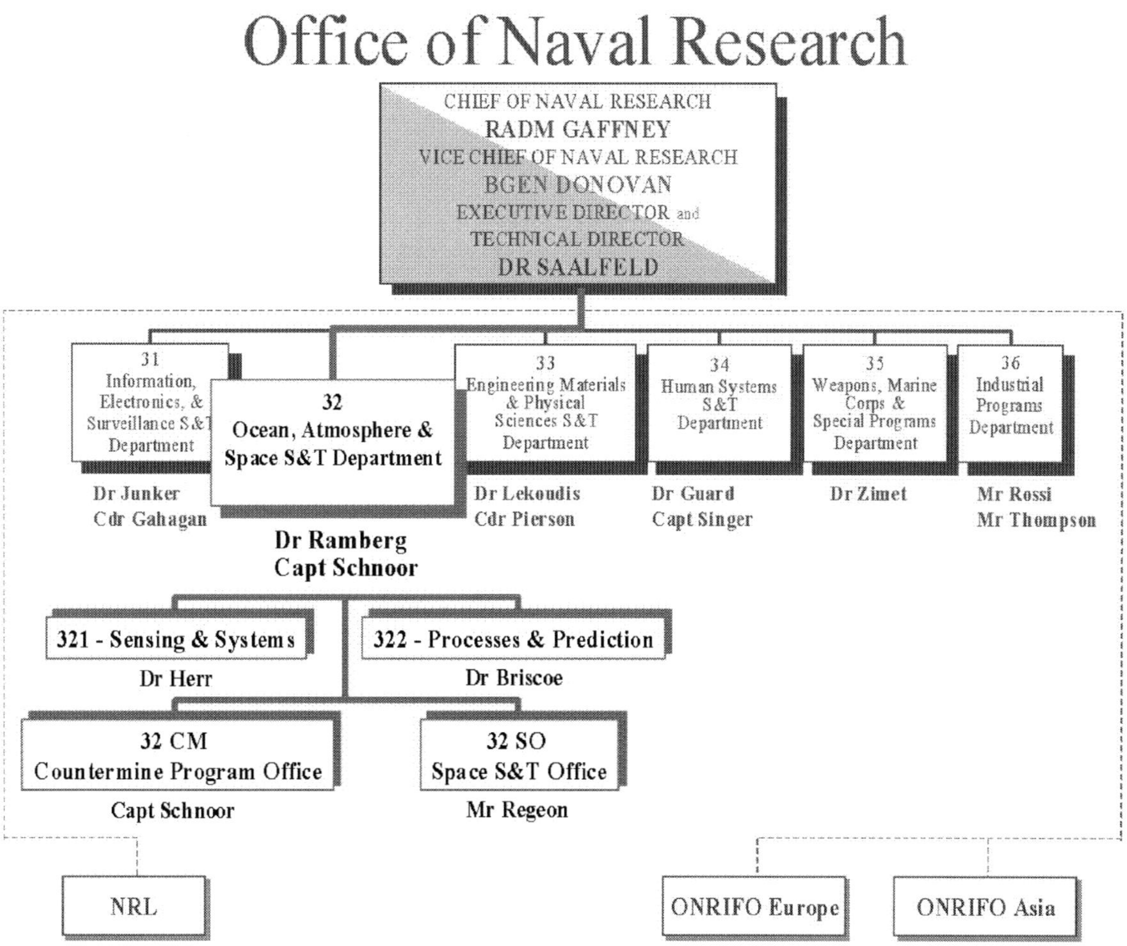

FIGURE 3-6 Organizational chart of the Office of Naval Research (ONR).

organizations. In addition, ONR provides technical advice to the operational branches of the Navy through the Chief of Naval Operations and the Secretary of the Navy (ONR, 1999).

The science and technology arm of ONR maintains close links with the oceanographic research and technical development communities by supporting proposals that will benefit naval operations (ONR, 1999). A major ONR goal is to encourage the transition of basic and applied research from the laboratory to the fleet. To facilitate this transition, ONR divisions use federal funding categories: 6.1 (basic and applied research), 6.2 (exploratory development), and 6.3 (advanced development). ONR also uses a series of program codes to identify areas of research focus (Figure 3-6). Most oceanographic research is funded in the Ocean, Atmosphere, and Space Science Technology Department (ONR 32) that also oversees the mine warfare science and technology program (Box 3-4). This program focuses on research that will improve the detection, identification, and neutralization of mines in open ocean and littoral environments while also improving offensive mining capabilities (ONR, 1999). Details of ONR research and technology development programs of special interest to mine warfare can be found in Appendix E.

BOX 3-4 ONR Department of Ocean, Atmosphere, and Space Science Technology Research Interests Relevant to Mine Warfare

- Develop a detailed understanding of littoral and deep water ocean and atmospheric processes that impact naval operations;
- Develop real time, environmentally adaptive sensors, processing, systems, and strategies;
- Develop distributed and autonomous ocean systems and methods to assimilate the large quantities of data they collect;
- Use multispectral and broadband sensors (undersea and remote) for mine warfare;
- Assimilate multiple and varied data types into predictive models and simulations;
- Develop scientific approaches that do not require complete or perfect knowledge to model and simulate complex environments;
- Develop methods of ground truthing model and simulations;
- Develop lightweight, broadband, acoustic sources and volumetric arrays;
- Continue the development of technologies for ocean and atmosphere remote-sensing;
- Develop new classification algorithms to enhance the capabilities of open architecture systems;
- Develop automated target recognition signal processing for highly cluttered undersea environments. (ONR, 1999)

4

Mine Warfare and the Ocean Environment: Mission Influence, Research Issues, and Future Exploitation

A number of environmental parameters can have strong impacts on mine warfare operations. These factors, discussed below, can affect operations directly, as with the influence of currents on diver operations and sediment transport, or they can influence sensor performance, as is the case for water column and seafloor acoustic properties. The surf zone, which has few similarities with other offshore regions in either its dynamics or in operational strategies, will be discussed separately.

OCEANOGRAPHIC EQUIVALENCE OF NAVAL DEPTH RANGES

Naval operations in the coastal zone are divided into the following categories: surf zone (0-10 ft.), very shallow water (10-40 ft.), shallow water (40-200 ft.), and deep water (>200 ft.; Figure 1-2). These depth ranges are somewhat equivalent to regimes defined by oceanographers (Table 4-1). Oceanographic depth ranges are primarily based on physical oceanographic principles, although acoustic, biologic, and geologic characteristics are similarly distinct.

The surf zone is defined by oceanographers as the region where surface waves actively dissipate energy due to depth-limited wave breaking. The 10-ft. depth cutoff selected by the Navy corresponds to a typical global mean value, although oceanographers allow the boundary to vary with wave height and breaker type (an undesirable level of complexity for operational planning). The surf zone water column is always well mixed. Breaking waves make this region noisy and nearly opaque to both acoustics and optics. In the surf zone, sediment transport rates are high, and significantly affect mine burial.

The naval very shallow water regime corresponds to the oceanographically defined inner shelf (Table 4-1), a complicated region where the surface and bottom Ekman boundary layers merge, stratification can be transient, buoyancy fluxes from rivers are commonly important, and fluid motions can be dominated variously by wave, tides, or low-frequency currents. Very shallow water is a complicated transition zone for both the Navy and the oceanographer.

The naval regimes of deep and shallow water (depths greater than 40 ft.) are grouped by oceanographers into a shelf and slope regime. Dynamically, this is the region in which surface and bottom Ekman boundary layers can form and are usually present. Motions are predominantly wind driven, with some buoyancy effects close to shore. In this region, surface waves provide only slight bottom stirring, and stratification significantly influences both circulation dynamics and acoustics. Sediment characteristics within this depth range will vary according to antecedent geology and fluid energy near the seafloor.

TABLE 4.1 Comparison of Naval and Oceanographic Depth Zones

Naval Operational Depth Zones	Definition* (ft.)	Oceanographic Depth Zones	Definition (m)
Surf zone	0-10	Surf zone	Zone of wave energy dissipation
Very shallow water	10-40	Inner shelf	0-30
Shallow water	40-200	Middle shelf	30-100
Deep water	>200	Outer shelf	100-130
		Slope	>130

*Note change of units between the naval and oceanographic depth zone columns

METEOROLOGICAL VARIABILITY

Mission Influence

The influence of meteorological variability on nearshore mine warfare can be either direct or secondary. Direct influences are primarily related to the effects of atmospheric conditions on sensor capabilities. For instance, broken cloud cover can yield confusing shadows for optical sensing, while solid cloud and rain can degrade both optical and acoustic-sensing performance. Little can be done about these effects, although their influence on sensors can be forecast.

Secondary influences are primarily related to the atmospheric driving of fluid motions that affect mine burial and countermeasures. At smaller scales, wave forcing by local winds will complicate diver operations and influence the rate of mine burial or scour. On larger scales, winds can drive shelf circulation and dramatically change local optical and acoustic properties of the water column. In the coastal zone, surface wind patterns can be strongly influenced by local coastal topography and can exhibit large diurnal changes. Similarly, the presence or absence of turbid outflow waters from nearby rivers and estuaries is usually strongly dependent on local atmospheric conditions.

Research Issues

Presently, naval operations rely on several atmospheric models for meteorological predictions. The Navy Operational Global Atmospheric Prediction System (NOGAPS) is the only global meteorological model operated by the Department of Defense. This model is specifically designed to deliver medium-range weather forecasts, but it also provides supporting information for nearly every application run at the Fleet Numerical Meteorology and Oceanography Center. Regional atmospheric prediction models include the Coupled Ocean/Atmosphere Mesoscale Prediction System (COAMPS) and the Third Generation Wave Model (3GWAM). COAMPS consists of a complete three-dimensional atmospheric data assimilation system that can be run simultaneously with one of two ocean models and can also be integrated with 3GWAM.

Although the models discussed above have advanced predictive capabilities, as is the case with all numerical models, continued enhancement and calibration is necessary to improve the quality of predictions. This is accomplished through inter-comparisons of model results between the Fleet Numerical Meteorology and Oceanography Center and the National Centers for Environmental Prediction. In open ocean areas, model predictions are robust, whereas in coastal regions landmass influences can result in the breakdown of model predictions.

Solutions

Further research and development of models for coastal meteorology will aid the prediction of meteorological factors affecting mine countermeasures. Finer scales than are common for synoptic meteorology have to be resolved in space and time. Proximity to the coast has a number of direct and indirect consequences. Coastal topography (for example, the coastal ranges of the U.S. west coast) can cause atmospheric edge waves with associated progressive cloud fronts. Longshore-directed winds will drive upwelling or downwelling, and abrupt changes in water column characteristics. Local thermal gradients, especially at the land-sea boundary, often push fog bands over the coast, greatly reducing visibility. In addition, local strong wind fields can generate complicated local seas that augment ocean swell.

BATHYMETRY

Spatial and temporal variations in water depth and seafloor profile can influence the location and height of breaking waves, the position and strength of surface currents, and the propagation of the tide into very shallow waters. In the surf zone, temporal changes in bathymetry can influence local dynamics in times as short as one day. In deeper waters, the fractional changes are smaller and slower, but can still easily be sufficient to cause mine burial. Bathymetry forms the boundary condition for all fluid motions. This influence can extend beyond the local region. For example, an offshore bank can cause focusing of waves onto a short section of beach that sits adjacent to quiet zones. Prior knowledge of these conditions is an important aid to operational planning.

Mission Influence

Bathymetry enters the minehunting problem as a boundary condition for wave and current motions through scour or burial (burial related to beach profile "activity") and through bottom clutter characteristics. The role of bathymetry as a boundary condition for nearshore waves and currents has received considerable attention, particularly with regard to planning and safety of special operations forces. Bathymetry measurements become more complicated in shallow water because a) the fluid dynamics are increasingly sensitive to small scale features (shelf scale motions average over larger regions and are partially isolated from bathymetry details by the boundary layer and stratification); b) temporal changes are increasingly important; and c) traditional survey methods are unusable.

Profile activity, and hence burial potential, varies strongly in the fluid environment with activity generally increasing with decreasing depth. In the deep and shallow water zone, burial occurs either on seeding or slowly thereafter. On the other hand, bathymetry in very shallow waters changes rapidly due to a wide spectrum of bedforms, and mines can quickly be scoured down and covered over.

Research Issues

The sensitivity of wave and flow fields to the details of regional and local bathymetry can be studied with models, although this approach has not received much attention. Thus, recommendations of bathymetry resolution for operational requirements are often ad hoc rather than model-based. Similarly, there is only limited knowledge on the "shelf life" of a set of bathymetric measurements (the period of time for which depth measurements at a location can still be considered valid). Clearly this is depth- and wave-climate dependent, but predictive capabilities are lacking.

Little is known about the climatology, variability, and importance of small-scale bedforms in shallow waters. These likely provide an important mechanism for mine burial. Bedforms also affect flows through bottom dissipation and act as a clutter against which mines must be detected. Knowledge of the presence and persistence of low-clutter regimes (no bedforms) would be useful. In addition, as bathymetry may be related to antecedent geology, information gathered from onshore sequences may be of use in predicting changes in seafloor character and bathymetry.

Solutions

Existing data that describe the natural variability of bathymetry at different depths, over different time scales, and in various oceanographic environments should be exploited. Little is known about smaller-scale bedforms, particularly meter-scale mega-ripples that greatly complicate the clutter problem. The dynamics of moveable (sand, for example) but low-clutter (flat bed) bottoms should be studied. Such studies should keep the mine burial problem in mind.

Automatic clutter characterization is a significant issue. Hunting of bottom mines, while usually carried out visually, is technically a feature extraction problem for which bottom clutter worsens the signal-to-noise ratio. Improvement of this ratio by optimum site selection (mentioned above) is one approach; however, considerable success has been achieved in automated clutter classification and anomaly identification (as an aid to visual analysis) through a variety of signal-processing approaches, including fractal description. These approaches are worth pursuing.

CURRENTS AND TIDES

Currents are long-time-scale fluid flows arising from a wide variety of processes, such as seawater density gradients, direct wind forcing, and tides. Meteorological and oceanographic analysis, prior to mine warfare activities in the coastal zone, must be sophisticated enough to understand the complexities of forces driving current flow, determine how these forces are interacting with each other, understand the time scale of variability driving current flow, and understand how this may affect the mission. Because of their importance, current and tidal data have long been a major component of the Naval Oceanographic Office (NAVOCEANO) products for mine warfare support.

Mission Influence

Tidally driven changes of sea surface elevation vary globally from negligible to tens of feet in very shallow waters. In the shallow and deep water zones, these elevation changes have little effect on minehunting operations or effectiveness. Tidal effects primarily influence mine warfare operations in very shallow water and the surf zone, although in the surf zone tidal currents are usually negligible compared to wave-driven flows. Near-surface mines may begin to broach at extremely low tides, providing a substantial increase in their remote sensing signature. Tidal currents can cause a dip that keeps a moored mine below the surface, can increase scour of bottom mines, and can cause significant transport of drifting mines. Outside the surf zone, tidal currents can often exceed 1 kt and thus affect diver and marine mammal operations. In addition, the onset of wave breaking, with its increased complexity for minehunting and swimmer operations, depends on depth. Thus, timing of swimmer operations in proper phase with the tide can substantially improve both safety and effectiveness.

Currents affect operations, safety, and potentially scour. There is a general decrease in dominant length and time scale of current motions with decreasing depth. Deep and shallow water flows are geostrophic and low frequency (and are likely to be predictable), while very shallow water and surf zone currents are more likely to be directly forced through wind, wave-driving forces, or buoyancy fluxes due to runoff or river outflow. Naval skill in current prediction decreases with decreasing depth (shallowest flows must be handled statistically or must be measured).

Research Issues

Tidal models are very accurate in deeper waters, however in very shallow water and the surf zone large errors may arise due to nonlinearities (finite height to depth); bottom friction; boundary effects, such as reflection; and local forcing (storm surge). These issues are most important in very shallow water, where the tides are largest and the minehunting problems most difficult.

Some aspects of current modeling over continental shelves are relatively well understood. For example, the alongshelf current can be well predicted for the case of steady wind forcing over an idealized continental shelf. Cross-shelf flows are much less well understood (and have formed the focus of the Coastal Ocean Processes, or CoOP, research program). Known weaknesses in understanding include the details by which wind blowing over the ocean transfers momentum through the surface boundary layer to force shelf flows and the details of the bottom boundary layer that retards those currents. While the upwelling processes have received considerable study, the corresponding downwelling dynamics are relatively unknown (not simply the opposite). Both processes can have important effects on optical and acoustic properties, aside from the importance of the currents themselves.

With the advent of satellite imaging, oceanographers became aware of the rich array of smaller-scale eddies on continental shelves. With currents that often exceed 50 cm/s and spatial scales on the order of 20 km, these features can have large impacts on operations, but are difficult to predict. On the U.S. West Coast, eddies are commonly produced by the interaction of alongshelf currents with capes and other coastal topography. Eddies can also be generated through shear instabilities and can propagate into coastal waters, as is the case with the Gulf Stream. Adequate prediction of a coastal eddy field requires improved numerical modeling coupled with assimilation of either in situ or remotely sensed data.

At even smaller scales, it has become increasingly apparent that a variety of internal waves are generated at or propagate onto the continental shelf. These can yield substantial anomalies to expected water properties and often have surface signatures that are easily detectable by remote-sensing. Best known are the soliton signatures seen in radar images. The possible utility of these signals, and the possibility that alteration of expected signatures by submerged vehicles could be detected, are not well explored.

In very shallow water, local buoyancy fluxes become dominant; depend on proximity to local rivers and estuaries, as well as local weather; and require substantial improvements in small-scale modeling capability. In the surf zone, currents are primarily wave driven and depend strongly on details of the bathymetry, particularly the presence of sand bars and of topographic channels through bars that can cause strong fixed-rip currents. Presently, models of the current field over a barred beach are primitive.

Solutions

Tidal models are becoming increasingly capable, however model success critically depends on the quality of bathymetric data available. As was found for special operations, there is a strong need for high-quality bathymetry data and increased capabilities to collect these data on denied coasts. Extension of present capabilities into shallower water using nested grids, nonlinear physics, and better ground truth data (perhaps from remote-sensing) for assimilation purposes should provide good tidal predictions for minehunting and special operations.

Improvements in current predictions will require progress on three fronts. At the most basic level, there are aspects of the fundamental physics (for example, the dynamics of bottom dissipation through a bottom boundary layer) that are poorly understood. Second, due to the expected complexity of coastal flows there is a need for improvements in modeling capability coupled with data assimilation. Third, there is need for increased availability of data for assimilation, either in the form of clandestine in situ sensors or through improvements in the types, accuracy, and availability of remote-sensing data.

OPTICAL PROPERTIES

Water clarity has always been important for diver missions, in terms of safety and ability to see well enough to detect, classify, and neutralize mines. With the development of modern optical sensors and appropriate platforms for deployment, optical methods of detection and classification have been added to the MCM toolbox. Water clarity is primarily determined by absorption and scattering by phytoplankton, detrital particles, suspended sediments, and dissolved organic material, as well as by the water itself. Scattering by particles blurs images of mines and other dangers, whereas absorption dampens the power of natural or artificial light sources for both divers and optical sensors. Transmission properties of coastal waters vary spectrally as functions of the relative concentrations of absorbers and scatterers. Transmission can be predicted if the spectral absorption and scattering coefficients are measured through radiative transfer models. For divers and sensors using the sun as a light source, underwater visibility is also a function of sun angle and atmospheric conditions (Mobley and Mobley, 1994).

The value of MCM optics has been increased by recent development of in-water sensors for direct measurement of inherent optical seawater properties (i.e., absorption and scattering coefficients), generation of new information on sources of variability in major absorbers and scatterers, and the evolution of radiative transfer models. Progress in understanding in-water optics directly benefits the interpretation of hyperspectral remotely sensed data collected from satellite and aircraft.

Mission Influence

With the exception of the surf zone, where breaking waves, bubbles, and suspended sediments create acute problems, optical properties are important to MCM in the very shallow, shallow, and deep water environments. In-water optics play a role in diver missions, in detection and classification of mines through video imagery and other optics-based techniques, and in interpretation of remotely sensed hyperspectral imagery. The latter has relevance to the determination of bathymetry through satellite remote-sensing.

Visibility for divers is a major issue for both safety and ability to detect mines and neutralize them. Divers need to know their visual depth limitations both vertically and horizontally in natural sunlight, their visibility range with an artificial light source either at night or when they are below the euphotic zone, and their vulnerability to visible detection from above the sea surface. The ability to obtain the detailed vertical structure of attenuation coefficients, rather than a water column average, provides divers with valuable information for mission planning. For example, a thin layer with radically different optical properties may serve as an optical haven or shield for avoiding detection by the enemy. Conversely, a thin layer may obscure the diver's ability to see the bottom in an otherwise clear water column. In environments where bioluminescence is present, an additional safety issue for divers is whether bioluminescent light will be observed from the surface or will be rapidly attenuated.

Light Detection and Ranging (LIDAR) systems are operated from helicopters, and cameras for underwater video imaging are incorporated into towed packages or autonomous platforms for minehunting and classification. Some of the issues relevant to divers also apply to cameras and laser detection systems. If the waters are turbid or highly absorbing, the optical signal will be rapidly degraded, limiting the detection distance between the optical sensor and target. Turbidity results in a decrease in the depth that can be effectively probed by helicopter systems and a decrease in the area that can be effectively hunted by towed systems. Knowledge of local inherent optical properties could provide reliability estimates on optical measurements, including bounds for different sizes and shapes of mines under different turbidity conditions and decision criteria for whether camera or LIDAR measurements are worth collecting under low-visibility conditions. Information on vertical structure of optical properties could provide guidance for deployment strategies of in-water sensors.

In coastal regions, the optical environment tends to be vertically and horizontally variable. Causes of spatial variability include phytoplankton patchiness, tidal and estuarine circulation patterns, and topographically forced mixing that redistributes absorbers and scatterers in the water column and resuspends sediments. Some of the mechanisms that produce spatial variability in optical properties of the littoral zone are also responsible for generating temporal variability. Optical remote-sensing with sufficiently high-spatial resolution can provide a context for extending local measurements of inherent optical properties to larger scales in the coastal zone.

Research Issues

Inherent optical properties vary in time and space. In coastal environments, it is often difficult to separate temporal and spatial variability, because both are greatly influenced by high-frequency forcings, such as tides, local winds, and inherently unsteady turbulence. Advection from local circulation patterns plays a major role in optical variability. An optical profile at one time and location may be useless for predicting the range of the day for a broader region, or even for that location later in the day.

The three major contributors to optical variability in the coastal zone are phytoplankton, chromophoric dissolved organic material (CDOM), and suspended sediments. One part of the problem is to understand the factors responsible for changing the specific optical properties of phytoplankton and CDOM, such as photoadaptation of pigment concentration and type for phytoplankton and bleaching for CDOM. Another part is to determine how concentrations of the three contributors change by processes such as phytoplankton growth and death; CDOM in situ production, leaching from the sediments, input with freshwater runoff, or seepage; and sediment resuspension, floculation, and settling. A third part of the problem is to understand how these time-dependent processes are coupled with local circulation and mixing. Both measurement and modeling approaches are needed to better understand how optics change in the compressed nearshore space and time continuum.

Remote-sensing provides a comprehensive spatial picture of the surface optical properties, but also presents three major challenges. One is to develop algorithms for maximal extraction of information from hyperspectral ocean color remote-sensing, such as sediment concentration. The second challenge is to couple surface imagery with information on vertical structure, collected from a limited subset of locations, to produce a three-dimensional picture of water clarity. The third challenge is to provide spatial continuity during cloudy periods and between repeat satellite passes or aircraft overflights. Autonomous vehicles equipped to carry optical sensors can augment above-water remote-sensing measurements by providing vertical information and by providing coverage when above-water sensors cannot.

Solutions

Minehunters need to be able to turn data rapidly into knowledge. Massive matrices of data on inherent optical properties of the water column can be collected in a very short time, but processing and interpretation can be a challenge. What the diver wants to know is what can be seen vertically and horizontally with the naked eye. By knowing the spectral nature of attenuation in advance, the diver will be prepared for color aberrations. What the operator of a towed package for optical hunting and classification wants to know is the optimal deployment mode and an estimate of reliability of the data to be collected. User-friendly robust algorithms are needed to provide the divers and system operators with information in real time or near real time.

Absorption and scattering degrade the quality of images collected by video and digital cameras. Improved pattern-recognition algorithms could increase the reliability of the information extracted from degraded signals. By coupling multiple sensors, such as optical and acoustical, the reliability of mine detection or classification could be increased. High-frequency acoustics and optics might be combined to classify and size the particles causing scattering. Analysis of underwater images is time consuming and requires an experienced operator. Improved automated pattern-recognition programs could decrease the time between image collection and decision making.

A high-degree of spatial and temporal variability of optical properties is characteristic of many nearshore areas. In a rapidly changing environment, the concept of determining the range of the day from one or a few profiles is futile. An integrated program of measurement and data synthesis incorporating optical data from profilers on surface vessels or easily deployed mini-moorings, pods of autonomous vehicles that can be launched simultaneously, and satellites is necessary to provide a three-dimensional view of the environment. Much progress has been made in the last five years in developing sensors that measure inherent optical properties. In the near future, these sensors need to be incorporated into autonomous vehicles and easily deployable profiling moorings. Process models for predicting changes in optically active constituents need to be coupled with physical circulation models to provide a temporally updated three-dimensional view.

Archived databases of satellite remotely sensed data on ocean color can be mined to provide climatological estimates of scales of spatial and temporal variability for specific locales. Time scales of interest need to span the tidal cycle (mixing and advection), diel cycle (phytoplankton growth, diel winds), and event scales (wind and storms) to the seasonal and interannual scales. With the recent and planned launches of a variety of ocean color satellites by the United States and other nations, unprecedented temporal and spatial coverage of many coastal regions will become available. New methods of integrating data from different sensors should be developed and diverse methods of statistical analysis applied to extract meaningful information for MCM applications.

ACOUSTICS

The nearshore environment, because it is so dynamic, is acoustically complex, being characterized by high-reverberation, high-ambient noise, and volume and boundary micro- and macro-heterogeneities. High-frequency sonars, used for mine detection and localization, as well as for classification, are the primary tools of mine warfare. The frequency range of sonar systems in common use is reasonably broad, but is generally higher than for anti-submarine warfare. Sonar systems are expected to operate in the very shallow, shallow, and deep water zones, and they must perform through the entire water column, including near surface, seafloor, and sub-seafloor. A general problem faced is the presence of reverberative and low signal-to-noise and highly cluttered acoustic backgrounds. Furthermore, incorrect interpretations of mine or mine-like objects can occur when using sound velocity profiles that are not representative of in situ conditions. Another difficulty with high-frequency sonar is the classic flashlight-in-space problem: a highly directive spatial sensor with insufficient time to completely survey the required volume comprising the minefield.

Mission Influence

Sweeping

Mines can be triggered by broadcasting acoustic signals, in both a broadband noise and a narrowband line structure, with frequencies similar to those emitted by propellers and machinery on-board the vessel. The target location can also be simulated by varying radiated signal strengths and by activating directional transmitters that appear to be at a specific position in the water column. This process can be further refined by acoustically simulating specific high-value targets (e.g., troop transports and carriers) or vessels with mine countermeasure (MCM) capabilities.

Hunting

A mine can be as small as a sphere 3 ft. in diameter or as large as a cylinder 21 in. in diameter and 10 to 20 ft. in length (Table 2-2). These dimensions define the detection resolution required for effective minehunting. It is a simple calculation to divide the water mass of interest into appropriately sized volume cells to estimate the time required to properly search the area for mines. Once this is accomplished, the next step involves separation of mine-like objects from objects that are mines. The latter need to be investigated more closely using divers or other sensors. As indicated above, the large amount of time involved in identifying and separating true mines from mine-like objects greatly impacts naval missions. This creates a dilemma for MCM forces, as they are forced to survey the potential minefield at a resolution that slows the rate of search so much that the larger naval mission must be delayed or cancelled.

Classification

Acoustic visibility is not the same as optical visibility. Acoustic images must be interpreted with considerably less information than a video image. With the coarser resolution of a sonar image, many natural objects may be identified initially as potential mines, thus requiring additional image processing or other means of eliminating objects as potential hazards.

Localization

Countering mines is normally a serial process. That is, objects must first be detected, classified, and visited with a remotely operated/autonomous vehicle (ROV/AUV) carrying a warhead. The warhead is then attached to (or near) the mine, the AUV moves away from the area, and the warhead is acoustically signaled to detonate. This process demands accurate navigational positioning during search and localization to reduce the time required for revisit and neutralization.

Research Issues

For convenience, research issues are grouped into three sections: (1) environmental parameters that impact acoustic performance, (2) acoustic capability, and (3) ocean acoustic models that are the basis for performance prediction.

Water column temperature and salinity datasets typically are not collected in sufficient detail for sound speed calculations at the resolution required for mine localization. In addition, these physical parameters are not predictable over the necessary operational time scales. Generally, knowledge of seafloor physical properties (such as sound speed, velocity attenuation, bulk density, and sediment roughness) are insufficient for acoustic modeling. Data are often not collected at an adequate resolution to be useful for minehunting sonars. Also lacking is an understanding of nearshore geophysical processes needed for development of interpolation and extrapolation methods to be used in data-sparse regions.

High-frequency scattering of acoustic energy from randomly rough surfaces cannot be quantitatively modeled with sufficient accuracy for sonar system performance prediction. Similarly, acoustic properties of the near-surface water column, which may have a significant bubble content due to wind and wave interaction, are not well enough understood to allow predictions of sonar performance.

Search algorithms and classification tools for intricate shapes in complex backgrounds are currently limited. The ability to "see" mine-like objects at varying degrees of resolution and at multiple "look" aspects needs to be exploited with new signal-processing techniques.

Solutions

Spatial Resolution

Measurements of water column and seafloor acoustic properties must be made at scales of less than 1 m. The resulting datasets will enable predictive models to be developed that can be used to interpolate and extrapolate acoustic properties to regions where data are sparse. Acoustic measurements from a variety of nearshore regions are required. Data must be collected from the entire water column to a depth of at least 2 m into the sediments. These data will provide important information for the development of robust acoustic models.

Temporal Resolution

Temporal aspects of mine warfare are complicated by mission planning times that are typically on the order of days and weeks and dynamic nearshore environmental processes that occur on time scales of hours and days. Prediction of environmental acoustic properties for planning purposes must be consistent with these time lines. Thus, the acoustic environment must be forecast for time scales of less than hours to weeks or greater.

Signal Processing

Signal-processing techniques that can be employed in complex (multiparameter) data fields should be developed. Also, signal-processing algorithms capable of exploiting low signal-to-noise situations are required. The exploitation of detailed seafloor data currently collected (phase and amplitude from sidescan sonar) can be applied to mine-like target recognition methods.

SEAFLOOR BOTTOM TYPE

Mission Influence

In the coastal zone, knowledge of seafloor bottom type is vitally important for successful mine warfare operations. The seafloor and its physical, chemical, and magnetic properties can be important in all aspects of the mine warfare problem, for example,

(1) mine burial probability, a function of sediment properties, drives sweep or hunt tactical decisions;
(2) seafloor conductivity and water depth are key factors for determining magnetic sweep paths;
(3) bottom reflectivity is a factor in airborne LIDAR performance;
(4) bottom sediment characteristics are a key factor in sediment transport, which affects water clarity and mine burial;
(5) minehunting sonar performance is normally bottom reverberation limited;
(6) sediment properties determine shock wave propagation, a method for mine neutralization in the surf zone.

Fortunately, naval science and technology efforts are relatively strong with respect to mine warfare-specific seafloor studies. The Office of Naval Research (ONR) has a basic and applied research effort (6.1[1]; see Chapter 3, "Support for Mine Warfare: Environmental Science Programs Within the Office of Naval Research") with the objective of improving the performance and prediction of mine warfare systems used to detect, classify, and neutralize mines located within or on the seafloor. Furthermore, an exploratory development (6.2) mine warfare program is underway that is oriented toward addressing bottom backscattering issues at high-frequencies.

In the MCM research area, ONR's 6.1 High-Frequency Scattering program is investigating high-frequency scattering issues for various types of sediments in shallow water. ONR is also funding the 6.2 Real Time Swathmapper for rapidly mapping shallow areas using an unmanned undersea vehicle (UUV) platform and the Hyperspectral Classification of Coastal Environments, a remote-sensing effort designed to use multi-frequency techniques to classify seafloor sediments.

To better understand and quantify the significance of the mine warfare problem, the U.S. Navy has embarked on a course that includes coordination and analysis of worldwide seafloor data and development of mission planning systems, such as the Mine Warfare Environmental Decision Aid Library (MEDAL). MEDAL uses descriptive parameters to define bottom composition for a wide range of depositional environments. Information can be obtained in situ from diver reports, extracted from acoustical data, or viewed from a video camera on a mine neutralization vehicle. Bottom sediment databases containing 70 categories of sediment descriptions are automatically input into MEDAL. In MEDAL, there is a direct mapping of these descriptions into 15 categories for the sonar performance model and into mud, sand, and rock for the mine warfare (MIW) doctrine worksheets. Bottom roughness and clutter density databases are input into MEDAL as defined by MIW doctrine. MEDAL also provides manual entry for the above data types based on diver observations, hydrographic charts, etc. Even though detailed sedimentological data are often available from bottom grabs and dive reports, these data are decimated to mud, sand, or rock based on doctrine as given in NWP 3-15 (Chapter 2, "Mine Warfare Doctrine").

Operationally, a better understanding of seafloor bottom types and properties can lead to increased detection range and search rate, improved classification performance, and increased capabilities for buried mine detection. Thus, improved understanding of seafloor properties will greatly improve mine warfare mission efficiency.

Research Issues

Significant naval research efforts have been directed toward a better understanding of sediments and seafloor processes. These studies have demonstrated the importance of seafloor characteristics for littoral mine warfare, however significant deficiencies still exist, for example:

(1) A significant amount of acoustic energy has been observed entering the seafloor at angles below the critical angle. Since many acoustic mine warfare systems are operating near the seafloor, low-angle acoustic propagation is critically important to system performance. Work is required to fully understand the physics of this phenomenon.

(2) At this time, the use of distributed explosive arrays (e.g., explosive nets) appears to be the most effective way of neutralizing mines in the surf and landing craft zones. Although, it must be acknowledged that waves and wave-generated currents can cause significant deployment problems for explosive nets. Since most surf zone mines will be buried in the sediments, sedimentary explosive shock wave propagation is critical for determining operational performance. Presently, we do not understand the physics of sedimentary stress-strain relationships; therefore, we cannot adequately model or predict the effectiveness of the explosions for mine neutralization.

(3) Mine burial is an essential factor in our ability to conduct effective mining operations. If mines are buried in the objective area, we have no recourse but to conduct influence-sweeping operations. Therefore, to make this extremely important tactical decision, it is necessary to have the capability to predict the likelihood

[1]As discussed in Chapter 3, ONR divisions use federal funding categories: 6.1 (basic and applied research), 6.2 (exploratory development), and 6.3 (advanced development).

of mine burial. Presently, there are four mechanisms by which mines will bury: scour (current induced and wave induced), migrating sand ridges, burial by deposition, and impact burial. Existing models to predict mine burial are rudimentary at best. Of these four models, impact burial prediction is the most mature, but even this model needs improvement. We do not understand the forces causing mine burial (currents, sediment transport), in particular how these forces interact with various mine shapes. An understanding of these physical processes would allow an improvement of burial prediction models. A significant deficiency in this regard exists for the surf zone.

(4) Sediment transport in very shallow water and surf zones greatly affect mining operations. Sediment transport is critical for mine burial, electro-optic transmissivity, acoustic scattering, and bathymetry. Significant work has been done in sediment transport research, and a strong understanding of the physics involved has been developed, but these efforts have been generic rather than mine warfare specific.

Solutions

Research efforts should focus on advancing ongoing sediment acoustic property research in the MCM Tactical Environmental Data System (MTEDS) Program, which includes development of techniques enabling the AN/SQQ-32 minehunting sonar to "calibrate itself" for better sonar performance by determining seafloor reverberation statistics. In addition, MTEDS researchers have developed techniques using MCM vessel echo-sounders, in conjunction with the Naval Research Laboratory's Acoustic Seafloor Classification System, to determine sediment physical properties necessary for mine burial prediction. Research efforts should also focus on the problem of below-critical-angle acoustic partitioning. This effort should be cooperative between the ocean acoustics and geology and geophysics communities.

As seafloor type and onshore geology may be related, predictions of seafloor character in the Surf Zone and Very Shallow Water environments can be accomplished with some success by extrapolating known stratigraphy and geologic structure from the adjacent landmass. There is an abundance of information on nearshore terrestrial geology but, comparatively, there is little known about the geology of continental shelves. Thus, the ability to predict seafloor character would benefit from expansion of seafloor databases to include onshore stratigraphic and structural information.

Development of mine burial prediction models should be initiated as part of the exploratory development phase (6.2) of the Environmental Physics for Mine Warfare Program. This research will improve existing impact burial prediction models and develop effective models for other mine burial mechanisms. Mine burial in the surf zone and empirical testing of quantitative models should also be addressed.

Development of an explosive shock wave propagation model should be initiated as part of the Environmental Physics for MCM Program. This research initiative should encourage multidisciplinary interactions among explosive physicists, geophysicists, and mathematical modelers.

It is also recommended that the excellent basic and applied research (6.1) from the various sediment transport programs be advanced to the exploratory development phase (6.2). Sediment transport research should also investigate the effects of this phenomenon on specific mine warfare areas of interest, such as mine burial, E-O system performance, acoustic scattering, and bathymetry.

SURF ZONE

Definition of the Region

By doctrine, the surf zone encompasses depths less than 10 ft. (~ 3 m), however the oceanographic definition is the presence of breaking waves. The physical extent of the surf zone, the nature of the dynamics, and the complexity of operations in this area all depend somewhat on the regional framework geology. Coral or mud beaches present different operational constraints than sandy beaches (the primary focus of this discussion). Similarly, the horizontal extent of the surf zone depends on the slope of the beach and on tidal range varying from tens of meters (~30 ft.) for steep slopes to a kilometer (0.6 mi.) for flat dissipative beaches.

Several classification schemes exist to organize surf zone characteristics according to geological setting and dynamic state (Inman and Nordstrom, 1971; Wright et al., 1986). The following discussion deals with a generic intermediate sandy beach and should be extrapolated for other environments.

The Fluid and Sediment Environment

The fluid environment of the surf zone is energetic and dominated by waves or wave-driven currents. Wave heights are commonly 1 m or more (> 3 ft.). Waves are turbulent, and inject bubbles into the water column through spilling and especially plunging breakers. Currents can exceed 1 m/s (1.8 knots) and can be directed alongshore for obliquely approaching waves or offshore in strong rip currents. Tidal currents are not usually important, but tidal changes in sea surface elevation can shift the surf zone on- or offshore and can substantially change the location and intensity of breaking and the strength of both rip currents and longshore currents.

Bathymetry and the sediment environment are commonly complex and highly changeable. Sand bars introduce meter-high anomalies in the beach profile that trigger wave breaking and can induce strong circulation patterns. Gaps in bars funnel rip channels that reinforce the gaps to form a rich suite of morphologies. Moreover, in most environments, bathymetry changes rapidly with a nearshore survey usually having a useful "shelf life" of a week or so, but as short as a day under storm conditions. Shorter-scale bedforms form or change very quickly in this environment, although perhaps conveniently the seabed is usually flat in the highest energy regions.

Mine Warfare Strategies

Mine warfare in the surf zone is unlike any other region because (a) the environment is hostile; (b) acoustic and optical search tools are generally unusable; (c) burial or scour can be rapid and extensive; and (d) the mines can be cheap, numerous, and mobile. As a consequence, brute force minesweeping methods are most often used. The effect of the environment on mine warfare operations can best be described in terms of the planning and in-stride execution phase and then a follow-up phase.

Operation Planning and In-stride Execution Phase

Because surf zone conditions are so difficult, explosive nets and line charges are the primary method of clearing safe lanes. Thus, planning requirements are primarily concerned with explosive effectiveness (related to pore water gas content) and selection of preferred (safest) routes. For the purposes of this report, the latter is addressed with reference to the mine problem versus other operational and safety issues.

Both of the above concerns are related to sediment characteristics and mobility. Gas content depends on depositional sedimentary organic matter content and the degree of exposure and recent turn over of sediments, the latter being somewhat related to bathymetric variability. Similarity, it is likely that in the time between seeding a minefield in the surf zone and an amphibious operation, the field may undergo substantial change. Heavier mines may become either partially or completely buried through profile variability (sinking during erosion, then covering during subsequent accretion) or through scour. This process is facilitated through all modes of profile variability from the movement of bars to bedforms and can easily result in burials up to 1 m in depth (potentially rendering a mine ineffective or inert).

Similarly, smaller anti-personnel mines are known to be mobile under wave and current motion. Thus, a minefield evenly spread initially may slowly become non-uniform as mines concentrate in, for example, rip channels. Knowledge of this process and of the recent morphologic changes in a beach (from remote-sensing) could allow selection of the safest routes with low-expected mine density and/or deep burial (alternatively, rip channels may be preferred if buried mines are a major concern).

Optimum use of these effects will require study of mine burial and redistribution processes on ocean beaches. It is reasonable to expect that the safest mine routes may also turn out to be the safest routes for other operational purposes.

Follow-up Minehunting and Sweeping

Minehunting and sweeping in the surf zone is exceedingly difficult. The high levels of turbulence and bubble injection usually render the surf zone opaque to both optical and acoustic search techniques. Water clarity on sandy or mixed grain size environments is often less than 1 m (3 ft.), and decreases rapidly in the presence of fine-grain sediment and biologic components. While diver search is often difficult to impossible, conditions can improve depending on the influence of local river or estuary outflow (for example, the direction and extent of the Chesapeake Bay outflow on the regional environment) or on local winds driving upwelling or downwelling flows that might move turbid waters offshore. These conditions are usually predictable with models and can be incorporated in later clearance planning.

The large bubble density in wave breaking acts both as a source of acoustic noise and as an effective absorber of acoustic energy from active sources. The break point is effectively a blanket that allows no shoreward sonic penetration from offshore active sources. The poor acoustic environment is nearly always a problem, but it is particularly so under plunging breakers. At sites with an appreciable tide range, advantage may be taken in search planning of tidal shifts of the break point and locations of strongest wave breaking.

Environmental parameters in the surf zone have significant influences on diver and marine mammal operations, which are the same as those affecting Special Operations Forces, detailed in a separate Ocean Studies Board (OSB) study (NRC, 1997). Prediction and understanding of waves and currents are very important to diver safety and effectiveness. However, in follow-up overt operations, issues that affected diver observability, such as bioluminescence and aerosols, are less important.

5

Discussion and Summary

ROLE OF THE SCIENTIFIC COMMUNITY IN MINE WARFARE

The research community has an important role to play in support of mine warfare by improving the understanding of coastal oceanic processes, including the physics of light and sound propagation in nearshore areas; developing quantitative oceanographic models; and developing advanced environmental measurement capabilities. Advances in these areas will provide the warfighter with the capability to predict oceanic changes, process vast amounts of data, and output these data as parameters useful for mine warfare planning and tactical decision making. The academic and engineering communities can also enhance mine warfare operations through advances in database development and data transfer technologies. Advances in data management techniques will enable complicated datasets to be used effectively and efficiently while also providing data processing capabilities. Wireless data transfer will enable real time processing of in situ environmental data transferred from mine countermeasure (MCM) vessels to the Naval Oceanographic Office (NAVOCEANO). The research community can also support mine warfare operations through the development of advanced training techniques to increase the quality and efficiency of oceanographic training for mine warfare operators.

Continued dialogue between naval operators and academic scientists is essential to ensure that researchers are aware of the current needs and capabilities of the mine warfare community. This symposium provided a unique opportunity to facilitate interaction between these two groups and jointly identify specific research challenges. Immediate results were seen through increased and enhanced cooperation between symposium participants and mine warfare operators.

Broadening interactions among academic scientists, operational scientists, and members of the operational Navy is also an important role for the Office of Naval Research (ONR). Programs to facilitate this interaction include the Secretary of the Navy/Chief of Naval Operations Oceanographic Research Chair, ONR Institution Scholar, Naval Science Assistance Program, Postdoctoral Fellowship Program, Young Investigator Program, Summer Faculty Research Program, Faculty Sabbatical Leave Program, Multi-disciplinary Research Program of the University Research Initiative (URI), Defense University Research Instrumentation Program, and the Defense Experimental Program to Stimulate Competitive Research. Details of these programs can be found in Appendix F and on the ONR Web page (ONR, 1999).

ENVIRONMENTAL PARAMETERS AND THEIR IMPORTANCE TO MINE WARFARE

Oceanographic parameters having the greatest impact on mine warfare operations were discussed in Chapter 4, which also outlined specific research issues and proposed solutions arising from symposium and steering committee discussions. A theme of these solutions is the need for enhanced predictive modeling capabilities for a wide range of oceanic processes that will greatly enhance the warfighter's ability to turn vast amounts of oceanographic information into knowledge used to control battlefield operations. The development of accurate and reliable models will be difficult due to the complexity of nearshore environments. Consequently, research on the nature of coastal zone processes should be encouraged and the results of this research should be applied toward quantitative descriptions of shallow water oceanography. In addition, the usefulness of model and database outputs will be greatly enhanced by real-time, 3D, graphical representations of oceanographic data. This "virtual oceanography" will provide the mine warfighter with the ability to visualize and manipulate environmental data in a more intuitive manner than is currently possible. This capability will greatly aid in mine warfare operational planning.

Other important issues related to oceanographic data and mine warfare that were noted during the symposium and in steering committee discussions were:

- The physical interaction of sound and light with the water column and seafloor provides the basis for developing powerful mine hunting tools. Despite this, there is still much work to be done to increase the fundamental understanding of these interactions in the shallow-water environment. In particular, enhancing the ability to measure and understand the propagation of high frequency (5 - 500 kHz) sound in the sediments typically encountered in coastal regions will lead to technical advances in underwater sensing systems to enable better mine detection.
- There is a significant need for signal processing enhancements for both optic and acoustic data. These enhancements will help maximize the amount of environmental information obtained from the large optic and acoustic datasets currently being collected.
- The ability to predict seafloor character and bathymetry may, in many instances, be greatly aided by a consideration of the structure and stratigraphy of onshore sequences. The development of databases containing information on regional geology could prove useful in the prediction of bathymetry and seafloor type in the Surf Zone and Very Shallow Water settings.

TRAINING ISSUES

The theory and practice of mine warfare is often omitted from the educational experience of the active duty Navy. Furthermore, there has been limited development of minefield theory since the end of World War II. Minefield theory is concerned with processes that provide information on the preferred geometry and density of a minefield, taking into account a given range of environmental conditions, threat characteristics, mine capabilities, and MCM characteristics. To ensure continued U.S. Naval control of the coastal zone, it is important that mine warfare theory develop along with technical advances in mine warfare operations.

The maintenance of an adequate state of readiness in mine warfare demands integrated research activities, technological development, procurement, and training. Presently, a large portion of operational capabilities and experienced mine warfighters remain in the Naval Reserve Force. This not only creates logistical difficulties, as mine warfare expertise is often far from areas of conflict, but also makes it difficult to incorporate mine warfare operations into training exercises (Hartmann and Truver, 1991).

Mine warfare training should be considered equally with more traditional branches of naval warfare, as the ability of naval forces to operate freely in the coastal zone is dependent on their ability to rid the battlefield of emplaced mines. Because of the importance of oceanic variability for mine warfare operations, the Navy should be committed in its efforts to ensure that the mine warfighter has access to advanced environmental techniques and the training required to operate, interpret, and maintain this technology. As MCMs evolve from dedicated to a mix of organic and dedicated forces, MCM training must also evolve. This will ensure that both forces are able to use environmental information being collected to better control battlefield operations.

DOCTRINE

Current MCM doctrine has changed little since World War II despite rapid changes in the technological complexity of mine warfare operations. Presentations and discussions at the symposium and during the writing of this report noted the need for significant changes in current doctrine (NWP 3-15). Continued increases in the use of highly specialized and complex technology for mine warfare operations, and the greater availability and reliance on large environmental datasets, make it essential that mine warfare doctrine be continually adapted, evaluated, and altered to keep pace with these rapid changes.

CHALLENGES FOR MINE WARFARE

High environmental variability in the coastal zone imposes formidable challenges to the mine warfighter. Recent naval documents, such as "Concept for Future Naval Mine Countermeasures in Littoral Power Projection: A 21st Century Warfighting Concept" (Rhodes and Holder, 1998), have acknowledged the need for mine warfare forces to stay on the cutting edge of environmental data collection, assimilation, and processing. Characterizations of nearshore environments require the collection of multiple high-resolution datasets. Coastal variability also makes it difficult to develop and use predictive models for mine warfare mission planning.

Previous symposia in this series identified common shallow water oceanographic needs of other branches of the operational Navy, such as the capability to predict coastal atmospheric conditions and sea state for strike warfare (NRC 1992, 1996a,b) and the need to understand nearshore environmental dynamics for coastal anti-submarine warfare, amphibious operations, and special warfare (NRC, 1992, 1994, 1997). These common needs illustrate the wide-ranging benefits of basic and applied oceanographic research in areas relevant to mine warfare. Enhanced capabilities resulting from this research would benefit a broad spectrum of end users in the fleet.

SUMMARY OF OUTCOMES

Specific outcomes of symposium and steering committee discussions include:

- Training Issues – To ensure that oceanographic data collected by NAVOCEANO and MCM forces can be converted into knowledge useful for the mine warfighter, naval operational staff must receive appropriate training. This training should provide the warfighter with an in-depth understanding of oceanographic processes and the capability to interpret oceanographic data and determine environmental influences on mine warfare operations.
- Doctrine – Sedimentary parameters are currently the main input to mine warfare doctrine. Thus, battlefield decisions are made without the use of high-quality oceanographic datasets being collected by NAVOCEANO and MCM forces. To ensure that battlefield decisions accurately represent prevailing environmental conditions, current doctrine should be updated to use a wider range of environmental data and robust statistical information.
- Data Collection and Database Capabilities – The expansion of naval operational focus in shallow water environments necessitates that the naval meteorologic and oceanographic community continue to develop methods for rapid collection of high-resolution environmental data. These data are essential for characterization of the high-variability in these environments. High-volume data collection will require continued advances in database capabilities to enable production of oceanographic products tailored to mine warfare needs.
- Modeling Capabilities – Continued development and enhancement of quantitative predictive models to describe coastal zone oceanographic processes will provide the mine warfighter with enhanced mission planning capabilities. Modeling advancements will require a concerted research effort that focuses on needs specific to the Navy and will provide a significant challenge for shallow water oceanographic research.

There is little doubt that the capacity to project naval strength into the coastal zone is dependent on the ability to operate where mines may be present. Because of the importance of oceanic variations for mine warfare operations, the warfighter must be able to collect environmental data and compare these data to historical data; the tools and training must also be available to interpret and characterize the oceanic environment from this information. The sixth Ocean Studies Board (OSB) symposium on tactical oceanography successfully identified areas where advances in environmental science and technology will strengthen and enhance the Navy's mine warfare capabilities. The symposium also examined the present state of knowledge and the environmental predictive capability relevant to mine warfare. The outcomes of the symposium and the research will help the naval meteorological and oceanographic (METOC) and mine warfare communities improve U.S. mine warfare capabilities.

6

References

Abel, D. 1999. Navy reportedly does little to counter threat of mines. *Boston Globe* July 21: A20.

Avery, J. 1998. The Naval Mine Threat to U.S. Surface Forces. *Surface Warfare* May/June: 4-9.

Boorda, J. M., ADM, USN. Circa. 1996. "CSS Mine Countermeasures CNO White Paper: Mine Countermeasures - An Integral Part of Our Strategy and Our Forces." Web page, [accessed 9 July 1999]. Available at http://www.ncsc.navy.mil/CSS/Papers/cnopaper.htm.

Department of the Navy. NWP 3-15.41 - Mine Countermeasures (MCM) Planning and Procedures (General Instruction) (U). 1540 Gilbert Street, Norfolk, VA 23511-2785: Naval Doctrine Command. (Confidential Document).

Ellis, W. G. "Jerry," RADM, USN. 1998. CNO NO96 Briefing: The Navy's Tactical Use of Oceanographic Information. *SACLANT/EXPO '98 SYMPOSIUM: Emerging Maritime Imperatives for the Next Millennium.* Hosted by North Atlantic Treaty Organization (NATO); Supreme Allied Commander, Atlantic (SACLANT), September 4, 1998 in Lisbon, Portugal.

———. 1999. CNO NO96 Briefing: Network Centric Oceanography in USW. *Network Centric Undersea Warfare: A New Look at USW Teamwork.* Hosted by the National Defense Industrial Association, Undersea Warfare Division Spring Conference, March 2, 1999 in San Diego, California.

Hartmann, G. K. and S. C. Truver. 1991. Weapons that wait: Mine warfare in the U.S. Navy. Annapolis, MD: Naval Institute Press.

Inman, D. L. and C. E. Nordstrom. 1971. On the tectonic and morphological classification of coasts. *Journal of Geology* 79: 1-21.

Martin, D., CMDR, USN. 1997. Chaos in the Littorals. *Surface Warfare* 22, no. 4: 20-25.

Mine Warfare Command, Public Affairs Officer. 1999. "COMINEWARCOM Home Page." Web page, [accessed 7 September 1999]. Available at http://www.cnsl.spear.navy.mil/cmwc/.

Mobley, C. D., and C. D. Mobley. 1994. Light and Water: Radiative Transfer in Natural Waters. Academic Press. 592 pp.

Morison, S. L. 1995. Guide to Naval Mine Warfare. Arlington, VA: Pasha Publications.

National Research Council (NRC). 1991. Symposium on Tactical Oceanography. Washington, D.C.: National Academy Press.

———. 1992. Symposium on Naval Warfare and Coastal Oceanography. Washington, D.C.: National Academy Press.

———. 1994. Proceedings of the Symposium on Coastal Oceanography and Littoral Warfare. Washington, D.C.: National Academy Press.

———. 1996a. Expanding the Uses of Naval Ocean Science and Technology. Washington, D.C.: National Academy Press.

———. 1996b. Proceedings of the Symposium on Tactical Meterology and Oceanography: Support for Strike Warfare and Ship Self-Defense. Washington, D.C.: National Academy Press.

———. 1997. Oceanography and Naval Special Warfare: Opportunities and Challenges. Washington, D.C.: National Academy Press.

Naval Oceanographic Office (NAVOCEANO). 1999. Mine Warfare Campaign Plan: Route Survey Environmental Data Bases - Concept of Operations. *Technical Report (TR) 315.* Stennis Space Center, MS: U.S. Navy, 36.

Office of Naval Research (ONR), Public Affairs Office. 1999. "Office of Naval Research ." Web page, [accessed 27 July 1999]. Available at http://www.onr.navy.mil.

Rhodes, J. E., LT. General, USMC and G. S. Holder, RADM, USN. 1998. "Concept for Future Naval Mine Countermeasures in Littoral Power Projection: A 21st Century Warfighting Concept." Web page, [accessed 16 November 1999]. Available at http://192.156.102/mem.htm.

Watts, A. J., ed. 1999. Jane's Underwater Warfare Systems. 11th ed. Alexandria, Va.: Jane's Information Group, 259-76.

Wright, L. D., P. Nielsen, N. C. Shi, and J. H. List. 1986. Morphodynamics of a bar-trough surf zone. *Marine Geology* 70: 251-85.

Zwolski, M., LT. 1998. "Airborne Mine Countermeasures (AMCM) History." Web page, [accessed 15 October 1999]. Available at http://members.aol.com/helmineron/history.htm.

APPENDIXES

APPENDIX A

Steering Committee and Staff Biographies

COMMITTEE MEMBERS

David Bradley earned his Ph.D. from the Catholic University in 1970. His research interests include acoustics, marine geology, and geophysics. Dr. Bradley currently serves as a senior scientist and associate director, acoustics, at the Applied Physics Laboratory of Pennsylvania State University, and was a member of the Ocean Studies Board from 1996-1999.

Tony Clark received his Ph.D. from the University of North Carolina in 1974. His research interests include underwater acoustics, oceanography, and geophysics. Dr. Clark has been a professor in the Marine, Earth, and Atmospheric Sciences Department at North Carolina State University since 1996.

Robert Holman received his Ph.D. from Dalhousie University in 1979. His research interests include nearshore morphodynamics and nearshore physical oceanography. Dr. Holman has been a professor in the College of Oceanography at Oregon State University since 1993.

Mary Jane Perry received her Ph.D. from the Scripps Institution of Oceanography, University of California, San Diego in 1974. Dr. Perry is a professor at the Darling Marine Center at the University of Maine. Her research interests include a variety of direct and inverse methods to determine phytoplankton abundance using optical and remote-sensing techniques.

NATIONAL RESEARCH COUNCIL STAFF

Alexandra Isern (Study Director, from August 1999) joined the NRC Ocean Studies Board staff as a Program Officer on June 1,1999. She received her Ph.D. in Marine Geology from the Swiss Federal Institute of Technology in 1993. Dr. Isern was a lecturer in Oceanography and Geology at the University of Sydney, Australia from 1994 to 1999. During this time she was the Chair of the Ocean Drilling Program Australian Scientific Committee. Dr. Isern is the co-chief scientist for an Ocean Drilling Program cruise that will investigate the magnitudes of ancient sea level change (ODP Leg 194).

Dan Walker (Symposium Manager and Study Director, until July 1999) received his Ph.D. in geology from the University of Tennessee in 1990. He is currently a Senior Program Officer with the Ocean Studies Board of the National Research Council. Since joining the Ocean Studies Board in 1995, he has directed a number of studies including *Science for Decisionmaking: Coastal and Marine Geology at the U.S. Geological Survey (1999), Global Ocean Sciences: Toward an Integrated Approach (1998),* and *The Global Ocean Observing System: Users, Benefits, and Priorities (1997).* A former member of the both the Kentucky and North Carolina state geologic surveys, Dr. Walker's interests focus on the value of environmental information for policymaking at local, state, and national levels.

Shari Maguire (Research Assistant, from May 1999) received her B.A. from Miami University in 1994. She currently serves as a research assistant with the Ocean Studies Board. Ms. Maguire is studying biological sciences at the University of Maryland in preparation for medical school.

Jennifer Wright (Senior Project Assistant, until April 1999) studied fine arts for a number of years at several institutions, most notably the Corcoran School of Art in Washington, DC. A native Washingtonian, Ms. Wright resides in Riverdale, Maryland. She is now at the University of Maryland Biotechnology Institute, where she is the assistant to the Vice President for Administration and Finance.

APPENDIX B

Symposium Program

MONDAY, FEBRUARY 8

Symposium registration and security check-in
Bachelor Officers Quarters (BOQ) Conference Center

1600 - 2000: Security Check-in and Registration
1700 - 1900: Recorder and Facilitator Briefing: Explanation of note-taker's duties and overview of symposium wargame

TUESDAY, FEBRUARY 9

The symposium series was developed, in part, to facilitate interaction between the scientist and the warfighter. Day One of the symposium, therefore, emphasizes interaction with personnel of the Mine Warfare Command (COMINEWARCOM) and the Mine Warfare Training Center (MWTC). Small groups of attendees, accompanied by COMINEWARCOM personnel, will participate in a series of activities and tour relevant ships and facilities in an effort to increase the attendees' awareness of the challenges involved in mine warfare.

0730 - 0800: BREAKFAST
Corpus Christi Bay Club

PLENARY SESSION I: Welcome and Overview
Corpus Christi Bay Club

0800 - 0815: Welcome and Administrative Remarks:
Dr. David Bradley (Chair, Symposium Steering Committee)
RADM Dennis R. Conley (Commander, Mine Warfare Command)
Dr. Dan Walker (Symposium Manager)
0815 - 0930: Introductory Briefings:

0815 - 0835 Mine Warfare Overview
CDR John Brown (COMINEWARCOM)

0835 - 0845	Office of the Oceanographer of the Navy (N096) CDR Frank Garcia (N096)
0845 - 0905	Mine Warfare and Oceanography LCDR Mark Null (COMINEWARCOM)
0905 - 0920	Research to Support Mine Warfare Dr. Doug Todoroff (ONR)
0920 - 0930	Wrap Up

0945: Buses load and depart for helicopter hangar (HM 15) for displays and demonstrations

1045: Buses load and depart for Ingleside Naval Station

1130: Buses arrive Ingleside

1130 - 1215: LUNCH
Mine Warfare Training Center Enlisted Galley

TOUR INGLESIDE MINE WARFARE FACILITY

Group ALPHA

1230 - 1315:	Tour Mine Warfare Training Center
1330 - 1415:	Tour *U.S.S. Inchon*
1430 - 1515:	Tour *U.S.S. Warrior*
1530 - 1615:	Tour *U.S.S. Falcon*

Group BRAVO

1230 - 1315:	Tour *U.S.S. Inchon*
1330 - 1415:	Tour Mine Warfare Training Center
1430 - 1515:	Tour *U.S.S. Falcon*
1530 - 1615:	Tour *U.S.S. Warrior*

Group CHARLIE

1230 - 1315:	Tour *U.S.S. Warrior*
1330 - 1415:	Tour *U.S.S. Falcon*
1430 - 1515:	Tour Mine Warfare Training Center
1530 - 1615:	Tour *U.S.S. Inchon*

Group DELTA

1230 - 1315:	Tour *U.S.S. Falcon*
1330 - 1415:	Tour *U.S.S. Warrior*
1430 - 1515:	Tour *U.S.S. Inchon*
1530 - 1615:	Tour Mine Warfare Training Center

1630: Depart for Corpus Christi Naval Air Station (NAS)

1700: Meeting adjourns for the day

1800-2000: ICEBREAKER
Corpus Christi Bay Club

WEDNESDAY, FEBRUARY 10

Day Two introduces the attendees to the mission needs of typical mine warfare operations. Small groups will participate in a single-sided, seminar-style wargame to obtain an understanding of the tactical use of environmental information and identify desired capabilities in the overall area of environmental understanding and prediction in support of mine warfare. Potential research and technology development to provide these capabilities will be proposed and discussed on Day Three.

0730 - 0800: BREAKFAST
Corpus Christi Bay Club

PLENARY SESSION II: Wargame Introduction
BOQ Conference Center

0800 - 0815: Overview of structure and goals of the Wargame (See detailed gamebook in back of this program)
CDR Frank Garcia (N096) and CDR Thomas Hodgson (Naval War College)

0815 - 0845: Mine Warfare Threat Brief (CLASSIFIED SECRET)
CDR Mike Shumaker (Office of Naval Intelligence)

0845 - 0900: Wargame Scenario Brief
CDR Frank Garcia (N096) and CDR Thomas Hodgson (Naval War College)

0900 - 1200: Concurrent Group Discussions
BOQ Conference Center

1200 - 1300: LUNCH
Corpus Christi Bay Club

1300 - 1500: Concurrent Group Discussions
(BOQ) Conference Center

1500 - 1600: EXTENDED BREAK

PLENARY SESSION III: Lessons Learned

1600 - 1700: Wargame Briefbacks
1600 - 1615: Group ALPHA
1615 - 1630: Group BAKER
1630 - 1645: Group CHARLIE
1645 - 1700: Group DELTA

1700 - 1800: INFORMAL RECEPTION
Corpus Christi Bay Club

1800 - 1900: DINNER
Corpus Christi Bay Club

THURSDAY, FEBRUARY 11

The final day of the symposium will focus on lessons learned and the identification of ways for the ocean science community to support mine warfare. Attendees will be reorganized into topical working groups to provide an opportunity for colleagues to participate in a dialogue directed at problem solving. By identifying research priorities germane to providing COMINEWARCOM with the needed capabilities identified during Day Two, the attendees can help shape the future of oceanographic and meteorological support for the U.S. Navy and the Mine Warfare Command.

0730 - 0800: BREAKFAST
Corpus Christi Bay Club

PLENARY SESSION IV: Finding Solutions
Corpus Christi Bay Club

0800 - 0830: Overview of day's activities. Charge to the Working Groups.

CONCURRENT WORKING GROUPS
BOQ Conference Center and *Corpus Christi Bay Club*

0845 - 1200: Working Group Sessions

1200 - 1300: LUNCH
Corpus Christi Bay Club

PLENARY SESSION V: Working Group Briefbacks
Corpus Christi Bay Club
1300 - 1700: Working group briefbacks/discussion

1700 - 1730: Closing Remarks

1730: Symposium Adjourns

WORKING GROUP ASSIGNMENTS

Group ALPHA

Beach, Reg (facilitator)
Boss, Stephen
Carder, Ken
Clark, Tony (steering committee)
Foster, Diane
Frailey, Lisa (facilitator)
Hanes, Dan
Lingsch, Steve
Maye, Pete
Schnoor, Tim
Swean, Tom
Tamburri, Mario
Tettelbach, Fred
Vinson, Phil
Webb, Benny
Weilert, Cynthia (recorder)
Weller, Bob
Wheless, Glenn
Widmayer, Ray

Group BRAVO

Bane, John
Bradley, Dave (steering committee)
Chotiros, Nick
Fry, Lee (recorder)
Fuller, Andy (facilitator)
Garcia, Frank
Glover, Linda
Griner, Joel
Hodgson, Tom
Holman, Rob (steering committee)
Kattawar, George
Keene, Dave
Kinder, Tom
Olds, Bob
Pettus, Eric
Rivenbark, Brian
Trowbridge, John
Williams, Kevin

Group CHARLIE

Anderson, Steve
Arnone, Bob
Bellingham, Jim
Boatman, Jerry
Cleveland, Joan
Cowles, Tim
Dugan, John
Gardner, Jim
Houtman, Bob
Jacobson, Randy
Jaffe, Jules
Lehr, Steve
Maffione, Bob
Martin, Dave (facilitator)
Null, Mark
Richardson, Mike
Scanlon, Randy
Weidemann, Alan
Wilson, Jim
Yoder, Tim (recorder)

Group DELTA

Baggeorer, Art
Blain, Cheryl Ann (recorder)
Brink, Ken
Briscoe, Mel
Drake, Tom
Gunderson, Chris (facilitator)
Haeger, Steve
Healy, Anthony
Lascara, Cathy
Lingsch, Bill
Paluszkiewicz, Terry
Perry, Mary Jane (steering committee)
Pfeil, Fred
Shanebrook, Kathy
Spinrad, Rick
Stanic, Steve
Tubridy, Lisa
Von Alt, Chris
Walker, Dan

Appendix C

Symposium Participants

Steven Anderson, Woods Hole Oceanographic Institution, Massachusetts
Robert Arnone, Naval Research Laboratory, Stennis Space Center, Mississippi
Arthur Baggeroer, Massachusetts Insitute of Technology, Cambridge
John Bane, University of North Carolina, Chapel Hill
Reginald Beach, Office of Naval Research, Arlington, Virginia
Jim Bellingham, Massachusetts Insitute of Technology, Cambridge
Cheryl Ann Blain, Naval Research Laboratory, Stennis Space Center, Mississippi
Jerry Boatman, Naval Research Laboratory, Stennis Space Center, Mississippi
Stephen K. Boss, University of Arkansas, Fayetteville
David Bradley, The Pennsylvania State University, University Park
Melbourne Briscoe, Office of Naval Research, Arlington, Virginia
Kendall Carder, University of South Florida, St. Petersburg
Nicholas P. Chotiros, University of Texas, Austin
Tony Clark, North Carolina State University, Raleigh
Joan Cleveland, Office of Naval Research, Arlington, Virginia
Cynthia Corbin Weilert, Coastal Systems Station, Panama City, Florida
Timothy Cowles, Oregon State University, Corvallis
Thomas G. Drake, North Carolina State University, Raleigh
John Dugan, Arete Associates, Arlington, Virginia
Diane Foster, Dalhousie University, Halifax, Nova Scotia, Canada
Lisa Frailey, U.S. Naval Observatory, Washington, D.C.
Leolan Fry, Jr., Coastal Systems Station, Panama City, Florida
Andrew B. Fuller, Mine Warfare Training Center, Ingleside, Texas
Frank Garcia, Office of the Oceanographer of the Navy, Washington, D.C.
James Gardner, U.S. Geological Survey, Menlo Park, California
Linda Glover, Office of the Oceanographer of the Navy, Washington, D.C.
Joel Griner, Jr., PED Mine Warfare, Arlington, Virginia
Chris Gunderson, Naval Pacific Meteorology and Oceanography Facility, San Diego, California
Steve Haeger, Naval Oceanographic Office, Stennis Space Center, Mississippi

Daniel M. Hanes, University of Florida, Gainesville
Anthony Healey, Naval Postgraduate School, Monterey, California
Thomas Hodgson, U.S. Naval War College, Newport, Rhode Island
Robert Holman, Oregon State University, Corvallis
Bob Houtman, U.S. Naval Observatory, Washington, D.C.
David Howell, Office of Naval Research, Arlington, Virginia
Randall Jacobson, Office of Naval Research, Arlington, Virginia
George W. Kattawar, Texas A&M University, College Station
David Keene, Naval EOD Technology Division, Indian Head, Maryland
Tom Kinder, Office of Naval Research, Arlington, Virginia
Cathy M. Lascara, Old Dominion University, Norfolk, Virginia
Steven Lehr, Mine Warfare and EOD Branch (N852), Washington, D.C.
William Lingsch, Naval Oceanographic Office, Stennis Space Center, Mississippi
Steve Lingsch, Naval Oceanographic Office, Stennis Space Center, Mississippi
David Markham, Naval Research Laboratory, Washington, D.C.
David G. McFadden, Office of Naval Intelligence, Washington, D.C.
Mark Null, COMINEWARCOM, Corpus Christi, Texas
Bob Olds, SPAWAR Systems Center, San Diego, California
Theresa Paluszkiewicz, Office of Naval Research, Arlington, Virginia
Mary Jane Perry, University of Washington, Seattle, Washington (now at: Darling Marine Center, University of Maine)
Eric Pettus, Very Shallow Water Mine Countermeasures Detail, San Diego, California
Fred Pfiel, Office of Naval Research, Arlington, Virginia
Mike Richardson, Naval Research Laboratory, Stennis Space Center, Mississippi
Brian Rivenbark, Mine Warfare Training Center, Ingleside, Texas
Randy Scanlon, Naval Warfare Development Command, Norfolk, Virginia
Robert T. Schnoor, Office of Naval Research, Arlington, Virginia
Kathy Shanebrook, U.S. Navy Chief of Operations, Washington, D.C.
Mike Shumaker, Office of Naval Intelligence, Washington, D.C.
Peggy Skiles, National Security Office, National Research Council, Washington, D.C.
Richard Spinrad, Office of the Oceanographer of the Navy, Washington, D.C.
Steve Stanic, Naval Research Laboratory, Stennis Space Center, Mississippi
Thomas Swean, Office of Naval Research, Arlington, Virginia
Mario N. Tamburri, Monterey Bay Aquarium Research Institute, Moss Landing, California
Frederick Tettelbach, ASSTSECNAV (RDA), Washington, D.C.
Doug Todoroff, Office of Naval Research, Arlington, Virginia
John H. Trowbridge, Woods Hole Oceanographic Institution, Massachusetts
Lisa Tubridy, Naval Surface Warfare Center, Panama City, Florida
Phil Vinson, Office of the Oceanographer of the Navy, Washington, D.C.
Christopher Von Alt, Woods Hole Oceanographic Institution, Massachusetts
Dan Walker, Ocean Studies Board, National Research Council, Washington, D.C.
Ben Webb, NWC War Game Department - METOC Support, Newport, Rhode Island
Alan Weidemann, Naval Research Laboratory, Stennis Space Center, Mississippi
Robert Weller, Woods Hole Oceanographic Institution, Massachusetts
Glen H. Wheless, Old Dominion University, Norfolk, Virginia
Raymond Widmayer, Mine Warfare and EOD Branch (N852T), Washington, D.C.
Kevin Williams, University of Washington, Seattle, Washington
James H. Wilson, Naval Postgraduate School, Monterey, California
Jennifer Wright, Ocean Studies Board, National Research Council, Washington, D.C.
Timothy J. Yoder, Naval Research Laboratory, Washington, D.C.

Appendix D

Wargame Gamebook

CONCEPT OF GAME PLAY

The second day of the Tactical Oceanography Symposium allows participants to examine the Mine Warfare force of 2005 as it evolves to include an organic force. This will be accomplished by conducting a seminar style Wargame. Seminar style wargames provide a background scenario and an opportunity to examine relevant courses of action. The seminar groups will be facilitated in their examination by an experienced operator/scientist. From this examination, participants will identify the principal tactical oceanography capabilities required for the fleet in the 2010 timeframe. Potential technologies to provide these capabilities will be proposed and discussed on the following day.

The following is a summarized sequence of events for 10 February 1999.

OFFICE of Naval intelligence (ONI) Threat Brief: A Secret-level brief to provide a look at the current and projected mining threat.

SCENARIO BRIEF: A brief to provide the scenario and a basic concept of operations in which a Joint Task Force (JTF) staff faces a projected mine threat in conducting expeditionary operations in a littoral region.

GROUP DISCUSSIONS:

- Organization: The players will be organized into four groups reflecting depths of mine threat. Personnel will be assigned to groups to leverage either fleet experience or current assignment dealing with these areas or professional expertise in technologies or programs most applicable to providing capabilities in that zone. Groups will be as follows:

  - Surf Zone (High-water mark to 10 feet)
  - Very Shallow Water (10 feet to 40 feet)
  - Shallow Water (40 feet to 200 feet)
  - Deep Water (200 feet and below)

TASKING:

• During day two the players are acting as members of a Mine Counter Measures (MCM) Commanders staff. Their mission is to commence planning to enable the JTF Commander to place forces ashore. In working through the planning process, each group will concentrate on their assigned depth zone. At the end of the day each group will present a brief highlighting the impact of 2005 METOC capabilities on the MCM mission and proposing those capabilities necessary for the warfighter to better carry out his mission. These required capabilities would be the focus of day three discussions investigating technologies to meet them.

SCENARIO

Introduction

The Island of Nordica is located in the northwest Atlantic Ocean off the coast of Canada and southwest of Greenland. It is compromised of two sovereign states: the Popular Republic of Nordica in the northern half of the island and the Sasenaci Federation in the southern half. The latter federation is essentially defunct, and two regional states remain; Vestenland in the southwest portion of the island, and Estmark in the southeastern area. A civil war erupted between factions in Estmark and Vestenland in 2003. The United Nations brokered a cease-fire in 2004 and inserted an observer force and other agencies into Stephenville, Vestenland to facilitate a program of normalization and rehabilitation (social and economic) in both regions.

Following elections in July 2004, however, considerable strain has developed. In late August, a UN report on the situation to the Secretary General, in which she assessed that Nordica and Estmark intend to use force to coerce Vestenland authorities into revoking permission for UN presence, thus ruining the rehabilitation program. If necessary, force may be used to replace Vestenland authorities with puppets controlled by Estmark criminal interests. A Security Council resolution followed in early October, requesting NATO assistance and concurring in the preparation of a contingency operation to be activated at the UN's request.

In mid-October, 2004, the North Atlantic Treaty Organization responded positively to the UN request and commenced planning. The United States agreed to lead the military effort. On 7 December 2004, Joint Chief's issued a warning order designating the Commander of the Joint Task Force, the forces, and direction to commence operational level planning for Operation Rockhound.

Joint Task Force Commander Campaign Plan

The following information is extracted from the Joint Task Force Commander's Campaign plan for Operation Rockhound to facilitate planning.

Friendly Forces

Task Force 301 (Carrier Battle Group)
USS Washington (CVN-73) (Flag)
USS Vicksburg (CG-69)
USS Philippine Sea (CG-58)
USS Carney (DDG-64)
USS Zerr (DDG-84)
USS Ross (DDG-71)
USS Connecticut (SSN-22)
USS Springfield (SSN-761)

Task Force 302 (Mine Counter Measure Force)
USS Inchon (Flag)(Commander, MCM Squadron 2)
USS Ardent

USS Warrior
HM-14
USN EOD DET
USN VSW DET

Task Force 303 (Amphibious Task Force)
USS Iwo Jima (Flag)
USS New Orleans
USS Shreveport
USS Gunston Hall
USS The Sullivans

Task Force 306 (Landing Force) (MEF))
II Marine Expeditionary Force (MEF) Forward
Task Group 306.2 (Ground Combat Forces)
2D MARINE REGT (-)
Task Group 306.3 (Combat Service Support Element)
Task Group 306.4 (Air Combat Element)
Marine Medium Helicopter Squadron 266 (Reinforced)

Mission

On order, OP ROCKHOUND forces will deploy to theater of operations and protect Vestenland from external interference to enable the UN to achieve its legal mandate.

Course of Action

The belligerents' most dangerous course of action to counter the UN mission would be a preemptive invasion of Vestenland by Nordica. Although this would involve high risk for Nordica, the JTF must plan for this contingency. Our course of action will be to to establish sea and air control and to commit a landing force to conduct extended operations ashore.

Commander's Intent

My intention is to plan and prepare to establish sea and air control and commit a landing force to conduct operations ashore. The extent of external interference expected in Vestenland is unknown. OP Rockhound forces shall initially deploy into theater with maritime forces as the main effort to aggressively demonstrate resolve and force capabilities. The Amphibious Commander shall be prepared to conduct an amphibious assault through St. Georges Bay to establish the landing force ashore, securing the port facilities and airfield in Stephenville. The Land Component Commander, once ashore, will constitute the main effort.

Belligerent Situation

Nordica maintains the only viable military force on the island. It possesses a comprehensive arms industry, which is capable of sustaining the national inventory of spare parts and ammunition. The armed forces are well equipped and efficient.

- Composition:

–Army- Field force consists of a brigade group of 7000 personnel in four battalions (one air mobile, one amphibious, two motorized infantry). This force can be split into a second brigade. Primary brigade group is dispersed and normally operates in smaller battle groups. Air lift and sealift is limited. Tactical helicopters are adequate for the air mobile battalion.

–Navy- 12 CG/DD/FF plus two diesel electric submarines. Nordica has been acquiring limited quantities of modern mines from the international market.

–Air Force- Modern reconnaissance, strike, and bomber aircraft. Capable of a strong first strike capability in the littoral but limited in ability to sustain an air campaign.

- Locations:

–Maritime- Naval bases are at St. John's, Placentia Bay and on the northern tip of Nordica at St. Anthony. All are approximately 300 nm from St. George's Bay.

–Air- Gander and St. John's are capable of handling fast jet traffic.

Joint Task Force Mine Counter Measures Forces

- Organic

–Three Airborne Mine Counter Measures (AMCM) configured H-60 helicopters. Embarked on USS Vicksburg, USS Philipine Sea, and USS The Sullivans. Each can be configured to carry the Shallow Water Influence Minesweep System (SWIMS), the Airborne Mine Neutralization System (AMNS), the AQS-20/X Advanced AMCM Sonar, the Airborne Laser Mine Detection System (ALMDS), or the Rapid Airborne Mine Clearance System (RAMICS.)

–One Long-Term Mine Reconnaissance System (LMRS) unit embarked on USS Connecticut.

–Three Remote Mine System (RMS) systems embarked on USS Carney, USS Zerr, and USS Ross.

–Shallow water Assault Breaching Explosive and Distributed Explosive Technology systems embarked on amphibious ships.

- Dedicated

–USS Ardent and USS Warrior with AN/AQS-32 Minehunting sonar and AN/SLQ-48(V) Mine Neutralization systems.

–HM-14 embarked on USS Inchon with MH-53E helicopters carrying AN/AQS-14A Minehunting sonar.

Nordican Mine Threats

- Up to 1400 sea mines of varying types are in the Nordican inventory. These include mostly KMD, MYAM, and KRAB moored variants. There are also known to be small numbers of modern bottom mines (MANTA/ SIGEEL, ROKAN) landing/invasion mines (PDM-1, PDM-2, PDM-3Ya), and limpet mines.

–The KMD is a mine resting on the ocean bottom which detonates upon influence of a passing target. It is laid in waters from 4 to 200 meters.

–The MyaM and KRAB are mines moored to the bottom which detonate upon contact. They are laid in waters from 3 to 50 meters.

–ROCKAN bottom mines are produced in Sweden and may be laid in waters from 5 to 100 meters.

–The Italian MANTA mine, along with the Iranian reproduction SIGEEL is a bottom mine which can be laid in waters from 2.5 to 100 meters.

- The Nordicans train to employ former Soviet mining tactics likely to employ the following tactics:

–A perimeter mine barrier of bottom and moored mines would extend across St. George's Bay from Cape St. George on the north to 5000 m west of St. David's on the south. This minefield would be to disrupt amphibious task force shipping formations and to delay the start of the amphibious assault.

–A main minefield of ROCKAN mines would be seeded by small patrol craft 5000 to 6000 m seaward of the Stephenville beach area. The mission of this minefield would be to deny maneuver room for naval gunfire support and the ship to landing craft evolution.

–A continuous barrier of PDM type mines and obstacles would be placed in the very shallow water and surf zones immediately in front of the airfield and beaches to deny the landing force access to the beach. Russian made amphibious transports are available to assist in laying these fields.

ISSUES FOR DISCUSSION (FACILITATORS)

During the discussions, players in each group should consider the role played by METOC in the following issues. Some thoughts for discussion include, but are not limited to:

- Battlespace Preparation and information management

  –Are we projected to have the capability to determine how many different bottom classifications are required? What capabilities are required to support environmentally adaptive sensors?

  –In stride implication. What capabilities are required to rapidly build awareness of METOC factors critical to effective mission planning and execution?

  –Information/Data sharing capabilities. What capabilities are required to collect data after entering the battlespace or remotely (in-stride).

  –NETCENTRIC implications in building a common tactical picture

- Mine Detection

  - Given a progression of sea mine threats from deep to shallow water identify capabilities that will enable avoidance or neutralization of the threat.

  - Discuss potential impacts of environmentally adaptive sensors and dynamic conditions driven search capabilities.

- Mine Classification

  –Do projected fleet systems have the capability to enable a high Search rate, low false alarm rate

- Identification

  –What capabilities will enable high-confidence level identification of mines in the broadest range of METOC conditions?

- Neutralization –

  –For Neutralization systems, are there any capabilities that will be adversely or favorably influenced by variable METOC conditions?

- Self Protection - Identify technological implications and required new capabilities to protect platforms against projected new sea mines which might be acoustically transparent, mobile, or otherwise novel.

GLOSSARY

ABS - Assault Breaching System

ALMDS - Airborne Laser Mine Detection System. For several years, the U.S. Navy has been evaluating electro-optics as a method of locating sea mines. Lasers have become more powerful and compact and their wavelengths more tunable. The use of a blue-green laser, which has a frequency compatible with seawater, allows a Light Detection and Ranging (LIDAR) system to provide accurate information on the characteristics of targets at various water depths. This technology will provide the Fleet self-protection when travelling through choke points and confined straits, as well as rapid reconnaissance of minefields in support of amphibious operations. The Airborne Laser Mine Detection System (ALMDS) is an electro-optics-based mine reconnaissance system that detects and localizes drifting/floating and shallow water moored mines from tactical helicopter platforms.

AMCM - Airborne Mine Countermeasures

AMNS - Airborne Mine Neutralization System. The Airborne Mine Neutralization System (AMNS) is an expendable, remotely-operated device that will be employed by H-60 helicopters to explosively neutralize proud (unburied) bottom, close-tethered moored, and volume sea mines that are impractical or unsafe to counter using existing minesweeping techniques. The system will have a day or night, shallow- and deep water capability. Prior to the neutralization mission, a minehunting sonar or electro-optic system will have accomplished mine detection, localization, and classification. The AMNS will be flown to the mine location where it will deploy its expendable neutralization device to reacquire and neutralize the threat mine.

AN/AQS-14 and AN/AQS-14A (Aviation Systems) - The AN/AQS-14, an active-controlled, helicopter-towed minehunting sonar, is currently used in MH-53E Sea Dragon helicopters. It is a multi-beam, side-looking sonar with electronic beam forming, all-range focusing, and an adaptive processor. The system consists of three parts: a stabilized underwater vehicle, electro-mechanical tow cable, and airborne electronic console. The 10.7-foot long underwater vehicle can be maintained at a fixed depth above the seafloor or below the surface, and the thin, coaxial cable is armored and nonmagnetic. Sonar information is presented on two continuous waterfall displays.

An upgrade to the AN/AQS-14 system, the AN/AQS-14A, modifies the airborne electronics from an analog to a digital system and increases the size of the operator's monitor. A Post Mission Analysis (PMA) station has been incorporated into the system for use with the contact tapes after the mission is complete to identify and classify mine-like contacts.

AN/AQS-20 - The AN/AQS-20 is a helicopter-towed minehunting sonar consisting of a Mission Control Display Subsystem, an AMCM Console Subsystem located in the helicopter, and a Towed Body Subsystem. The towed body includes side-looking, gap-filling, volume-searching, and forward-looking sonars. The AQS-20 will be effective against bottom and moored mines in both deep and shallow waters. It will provide an increase in area coverage rate in comparison to the current AQS-14 system and can provide single-pass detection of both bottom and moored mines.

AN/AQS-20/X Advanced AMCM Sonar - The AN/AQS-20/X will be a helicopter-towed minehunting sonar system containing an integrated electro-optic identification (EOID) device. An outgrowth of the AN/AQS-20 developmental minehunting sonar program, AQS-20/X will be compatible and fully integrated into the H-60 helicopter. The system will provide Aircraft Carrier Battle Groups (CVBGs) and Amphibious Ready Groups (ARGs) an organic capability for rapid detection, classification, localization, and identification of bottom, close-tethered, and volume mines. This capability will enable CVBGs and ARGs to transit or avoid mined areas in choke points and littoral areas with a high degree of self-protection.

AN/SLQ-48(V) Mine Neutralization System (MNS) (Surface Ship System) - an unmanned minehunting submersible, is standard on both the MCM 1 and MHC 51 classes. The vehicle obtains its power and guidance commands from the launching ship through a 3,500-foot umbilical cable. After a target is detected and classified by the ship's sonar, the MNS, which is initially directed by ship's sonar data, proceeds to the target at speeds up to six knots. The vehicle carries a small, high-definition sonar and an acoustic transponder that enables the vehicle to be tracked by the shipboard sonar. There is also a low-light-level television for examining the target, with illumination provided by on-board floodlights. Propulsion is provided by two 15-hp hydraulic motors and there are two horizontal and two vertical hydraulic thrusters for the exact positioning of ordnance to the target. Two consoles on-board the ship monitor and control the vehicle's operation. The MNS can destroy bottom mines by placing an explosive charge near the mine or by cutting the cable of moored mines, causing them to rise to the surface for subsequent neutralization or exploitation.

AN/SQQ-32 (Surface Ship System) - The AN/SQQ-32 is a variable-depth mine detection and classification sonar. It is standard on all of the *Osprey* (MHC 51)-class ships and is being backfitted in MCM 2-9. An upgrade to the SQQ-30, is better at discriminating between genuine mines and other objects, and is able to identify objects with near-picture quality. The SQQ-32 displays search and classification information simultaneously and independently, using separate search and classification transducers in a stable, variable-depth body. Multi-beam operation increases the sonar's search rate. The SQQ-32 can also be used from the hull in shallow water.

C4I - Command, Control, Communications, Computers, Intelligence

DP - Deterrence and Prevention

DW - Deep Water

DET - Distributed Explosive Technology. Distributed Explosive Technology is a 180-by-180-foot distributed explosive net system with a 200-foot standoff range that is launched from an LCAC by two Mk 22 Rocket Motors. The system is designed to provide a wide swath of clearance in the zero-to-three-foot deep portion of the Surf Zone. The system is designed for operation with SABRE in graduated clearing of an approach lane from the sea for amphibious landing craft. Explosive effectiveness tests have shown DET to be very capable of defeating mines in water depths from three to ten feet.

EOD - Explosive Ordnance Disposal

Explosive neutralization - The Explosive Neutralization program will incorporate enhanced explosive technologies, delivery systems deployable from LCACs, and advanced delivery systems deployable from unmanned, fixed-wing aircraft to provide for in-stride clearance of mines in the surf, craft-landing, and beach zones of the Amphibious Objective Area. The system will eliminate the need for time consuming manned clearance operations in hostile environments. The goal of the program is to maximize the effectiveness of wide area clearance against a defended beach, eliminating the need for lengthy manned clearance operations, thereby minimizing the loss of personnel and surface landing craft. The program has designed, fabricated, and demonstrated extended standoff line charges and nets. Also, an LCAC auto-pilot, known as SKIPPER, has been developed to increase system deployment accuracy and decrease mission time, thereby enhancing the capability of the basic SABRE and DET explosive systems.

LLSS - Laser Line Scan System. Currently, sonar minehunting systems cannot distinguish between actual mines and mine-like objects. Therefore, every mine-like contact must be investigated, resulting in very slow rates of identification and neutralization. The Laser Line Scan System (LLSS) program will provide a means of positively identifying actual mines, resulting in a projected mine neutralization phase only one-fifth as long as is currently required. This program will integrate the Northrop Grumman SM-2000 laser line scan sensor into the existing AN/AQS-14A sonar minehunting towed body.

LMRS - Long-term Mine Reconnaissance System (Organic System) The Long-term Mine Reconnaissance System ultimately will replace the Near-term Mine Reconnaissance System. The LMRS will operate from and interface with *Los Angeles* (SSN 688)-class submarines and the New Attack Submarine (NSSN). Like NMRS, LMRS will detect, classify, and localize mine-like objects. However, LMRS will provide improvements in areas where NMRS is limited, such as the overall number of systems, as well as its endurance and search rate.

MCM - Mine Countermeasures

MDA - Mine Danger Area

METOC - Meteorology and Oceanography

MIW - Mine Warfare

Mines - Quickstrike (US). The Quickstrike family of aircraft-laid, shallow water bottom mines is closely related to an earlier family of mines named Destructors. Quickstrike can use two variable-influence target detection devices to detect submarines and surface ships. The Target Detection Device (TDD) Mk 57 uses a magnetic/ seismic sensor. The Target Detection Device Mk 58 uses a magnetic/ seismic/pressure sensor. Both detection devices were approved for production in the early 1980s. The Quickstrike design emphasizes ease of maintenance, preparation, and use. The mines will either sterilize or self-destruct at the end of life. The Target Detection Device Mk 71 was approved for production in 1995 and uses a magnetic/seismic/pressure sensor. While the earlier devices use hard-wired algorithms with programmable sensitivities, this device is fully programmable for algorithms and sensitivities.

Mines - Quickstrike (US) Mine Mk 62 and Mine Mk 63. The Quickstrike Mine Mk 62 and Mk 63 are air-delivered bottom mines that use General Purpose Low-Drag Bombs Mk 82 and Mk 83 (500 and 1,000 pounds, respectively) as the explosive payload. Because a specialized kit is used to convert bombs into mines, demand for magazine space on aircraft carriers is dramatically reduced. These mines use either the Target Detection Device Mk 57 or, when available, the Target Detection Device Mk 71. Arming takes place at a pre-set time after the mine enters the water.

Mines - Quickstrike (US) Mine Mk 65. The Quickstrike Mine Mk 65 is a thin-wall, 2,000-pound air-delivered bottom mine. The mine may use either the Target Detection Device Mk 57, Target Detection Device Mk 58, or, when available, the Target Detection Device Mk 71. Arming takes place at a pre-set time after the mine enters the water.

Mines - Submarine Launched Mobile Mine (SLMM) (US) Mk 67

The Submarine Launched Mobile Mine (SLMM) Mk 67 is a modified Torpedo Mk 37 with a thin-wall mine warhead. It is delivered by submarine and is considered a clandestine mine. SLMM may use either the Target Detection Device Mk 57, or, when available, the Target Detection Device Mk 71. It is a shallow water, bottom mine that detects submarines and surface ships. The mine will either sterilize or self-destruct at the end of life. SLMM is launched from a submarine as a torpedo. After running to a pre-selected location, the torpedo motor shuts down and the weapon sinks to the bottom. Arming takes place at a pre-set time or distance after the torpedo run period.

Mines - Improved Submarine Launched Mobile Mine (ISLMM) (US)

The Improved Submarine Launched Mobile Mine is based on converting existing Mk48 torpedoes into mines. It features dual mine sections (warheads) to increase submarine mine laying capacity and has improved compatibility with the SSN-688 submarine fire control systems. In addition, ISLMM will have a multiple waypoint turn capability and greater range than the current SLMM, significantly increasing the number of minefields that can be planted by submarine.

Mines - Target Detection Device Mk 71

A mine Target Detection Device is the electronic fuse that observes changes in the environment in order to detect ships and submarines and decide whether the target is close enough to damage. Current bottom mines (Quickstrike and SLMM) use a TDD Mk 57 (TTD57), which is a magnetic/seismic device designed specifically for Cold War targets, such as large combatants and submarines on or near the surface. The Quickstrike Mk 65 can also use the TDD Mk 58 (TDD58), which is a magnetic/seismic/pressure device. The TDD57 and TDD58 are in the U.S. Navy's service inventory and are the only sensor/detection packages for bottom mines. The TDD Mk 71 (TDD71) is a programmable device capable of responding to emerging threats, such as quiet diesel-electric submarines, mini-submarines, fast patrol boats, and air cushioned vehicles. It adds an enhanced pressure sensor and has the capability to respond to remote control signals. The TDD71 is designed for use in all Quickstrike-series mines, SLMM, or Improved SLMM.

MMS - Marine Mammal System

MTW - Major Theater of War

NMRS - Near-term Mine Reconnaissance System (ORGANIC MCM SYSTEM). Both near-term and long-term programs have been established to provide the submarine community a UUV mine reconnaissance capability. The initial capability has been designated the Near-term Mine Reconnaissance System (NMRS). The single NMRS will be launched and recovered from a *Los Angeles* (SSN 688)-class submarine. NMRS will be capable of limited mine detection, classification, and localization with an inherent low-risk to the host platform. NMRS capitalizes on existing technologies and capabilities in order to reduce cost and become available in the near term. The NMRS will be in operation until the Long-term Mine Reconnaissance System (LMRS) reaches its Initial Operational Capability (IOC). NMRS is deployed through standard SSN-688-class torpedo tubes. The operational system will consist of two reusable UUVs; launch and recovery equipment; and shipboard control, processing, and monitoring equipment. Operators control the vehicle via a fiber-optic cable connected to the launch platform.

NOMBOS - Non-Mine, Mine-Like Bottom Objects

NWC - Naval War College

OMFTS - *Operational Maneuver from the Sea*

POR - Program of Record

PW - Propelled-Warhead Mines

RAMICS - Rapid Airborne Mine Clearance System. The Rapid Airborne Mine Clearance System (RAMICS) is a helicopter-borne weapon system that fires a special 20mm supercavitating projectile from a modified Gatling gun controlled by targeting algorithms and a blue-green LIDAR (light detection and ranging) to neutralize surface and near-surface mines. At the heart of this system is a supercavitating 20mm projectile that is specially designed for traveling tactical distances in air and water and driving a chemical initiator through a casing into the mine. The LIDAR locates and targets the mine and provides aiming coordinates to the gun's fire-control system. A burst of rounds is fired at the mine causing positive neutralization of the mine.

IPB - Intelligence Preparation of Battlespace

RMS - Remote Minehunting System (ORGANIC MCM SYSTEM) is a high-endurance, remotely operated, surface-ship launched and recovered semi-submersible vehicle towing a mine reconnaissance sonar. The system will conduct rapid reconnaissance of bottom and moored mines from the deep water region to the 30-foot contour of the very shallow water region. Mine reconnaissance operations conducted by RMS will determine the presence of mine-like objects and safe routes or operating areas around potential minefields.

SABRE - Shallow water Assault Breaching System. The Shallow water Assault Breaching System (SABRE) is a discontinuous line charge system delivered by the Mk 22 Mod 4 rocket and deployed from an LCAC. SABRE is designed to accomplish wide area neutralization of anti-invasion mines in the three-to-ten-foot deep portion of the Surf Zone, thereby minimizing the loss of personnel and surface landing craft. The system is designed for operation with the Distributed Explosive Technology (DET) in graduated clearing of an approach lane from the sea for amphibious landing craft. Explosive effectiveness tests have shown line charges to be very capable of reducing light and medium obstacles on land and defeating mines located in water depths up to three to ten feet.

SPM - Self-Protective Measures

SR - Surveillance and Reconnaissance

SWIMS - The Shallow Water Influence Minesweep System (SWIMS) is a self-contained system designed to perform high-speed magnetic or magnetic/acoustic influence minesweeping missions in a shallow water environment. SWIMS is fitted on MH-53E Sea Dragon helicopters, permitting full speed transit to and from the mission area, traveling over-the-horizon. It consists of a towed magnetic and acoustic source, a tow/power delivery cable, a power conditioning and control subsystem, and an external or palletized power supply. The magnetic device is smaller and lighter than other magnetic minesweeping sources, measuring ten feet in length, 20 inches in diameter, and weighing approximately 1,000 pounds. Its small size and reduced weight require minimum handling equipment, and it is deployable from the helicopter by two crewmen.

SWIMS is deployed from the helicopter by a standard tow cable when the helicopter reaches the area of operation. The programmable waveforms allow flexible performance and address emulation requirements. Installation uses existing AMCM interface equipment. SWIMS can be used in conjunction with the Mk 104 acoustic sweeping systems. The tow body is torpedo shaped and provides a hydrodynamically stable minesweeping platform that exhibits excellent towing characteristics under all operating conditions. SWIMS possesses high maneuverability in shallow waters and is capable of being towed at speeds up to 40 knots. Its ability to fully demagnetize allows the system to be transported within the helicopter allowing for fast transit to over-the-horizon operating areas.

SW - Shallow Water

SZ - Surf Zone

TDA - Tactical Decision Aid

USW - Undersea Warfare

UUV - Unmanned Underwater Vehicle

VSW - Very Shallow Water

VSW DET – Very Shallow Water MCM Detachment. Since 1996, the Navy has successfully completed a feasibility demonstration of the Very Shallow Water (VSW) MCM Test Detachment and authorized establishment of a permanent VSW MCM Detachment under the Commander, Mine Warfare Command. This detachment provides a capability for conducting advance force and pre-assault exploratory and reconnaissance MCM operations in the VSW zone (ten to 40 feet) and locating (and if necessary clearing) potential landing sites in support of joint littoral power projection operations.

The VSW MCM Detachment, comprised of EOD, Naval Special Warfare, and USMC Reconnaissance divers and marine mammals, is responsible for developing tactics, techniques, and procedures for MCM operations in the VSW zone and for rapidly mobilizing to embark deployed amphibious task groups during contingencies. The detachment is, in essence, the Navy's warfighting laboratory for evaluating prototype systems that could be used to conduct VSW MCM operations. Acquisition programs are underway to develop an Integrated Navigation Sonar System (INSS) and a hydrodynamic Underwater Breathing Apparatus (UBA) for VSW MCM divers. Additionally, the EX-8 Marine Mammal System acquisition program began in FY 98 to merge demonstrated prototype capabilities from the VSW MCM Test Detachment with new technology to acquire an operational MMS Det by the end of FY 00.

VSW MCM acquisition projects currently underway are only the beginning of a longer-term plan to acquire a viable VSW MCM capability that enables seamless power projection forward, from the sea through the littoral waters and across the shore, even when a credible mine threat challenges exists. The VSW MCM Detachment is the small cadre of experts that will remain focused on solving the VSW MCM problem and enable the transition from a dedicated contingency configuration to a CVBG/ARG-focused organic configuration and from a "diver-in-the-loop" configuration to an unmanned system configuration.

Appendix E

Environmental Science Programs Supporting Mine Warfare Within the Office of Naval Research

Sensing and Systems Division

OAS Science and Technology is divided into two divisions. The first is the Sensing and Systems Division, which manages research related to Navy and Marine Corps operations, including mine warfare. The following are summaries of research and technology development of special interest to mine warfare.

Ocean Acoustics

The Ocean Acoustics Program supports research that addresses the physics of generating, propagating, and scattering narrowband and broadband acoustic (and elastic) waves in the ocean environment. This program encompasses three primary research areas:

- Shallow water acoustics, which aims to understand the propagation and scattering of low-frequency (10Hz to a few kHz) acoustic energy in the nearshore ocean environment. Areas of research include investigations of shallow water scattering mechanisms, the conversion of seafloor-incident acoustic energy into elastic body waves and interface waves, and acoustic propagation through linear and nonlinear internal waves.
- High-frequency acoustics, which seeks to understand the interaction of high-frequency (few kHz to MHz) sound in the ocean environment, including the propagation of sound through an intervening turbulent or stochastic medium; scattering from rough surfaces, biologics, and bubbles; and penetration and propagation in the seafloor.
- Long-range propagation, whose goal is to understand the behavior of long-range sound propagation (several hundred kilometers to several thousand kilometers) in the ocean. The main area of study concerns the effect of internal ocean waves and mesoscale waves on transmitted broadband acoustic signals.

Remote-sensing

This research program investigates physical and chemical processes governing Earth surface scattering of active and passive electromagnetic radiation and the propagation of this radiation through the upper atmosphere and the near space environment. Remote-sensing research is directed toward development of clutter models and automatic target recognition. Additional research interests include studies of electromagnetic scattering theory, microwave properties, scattering surface characterization, and wave and flux modulation mechanisms.

Coastal Dynamics

Research in the Coastal Dynamics Program includes aspects of coastal ocean fluid and sediment mechanics. At present, three research areas are emphasized:

- Inner-shelf dynamics: fluid mechanics of the continental shelf, particularly the inner shelf, seaward of the surf zone, where surface and bottom boundary layers influence much of the water column;
- Nearshore processes: fluid dynamics, fluid-sediment interactions, and the resulting morphological response in the nearshore, where waves begin to break because of shoaling; and
- Surface waves: the fluid mechanics of coastal surface waves and methods for improved prediction.

New initiatives will explore bay, estuarine, and river-mouth dynamics, including the interaction of water depth, tides, winds, and water column stratification. Research and development collaborations with other programs help address such issues as coastal meteorology, coastal zone remote-sensing, shallow ocean modeling, and mine burial and migration. Particular interest is placed on the use of remote-sensing techniques that can be used to augment limited or minimal in situ field measurements.

Ocean Engineering and Marine Systems

This program focuses on multidisciplinary science and technology research on the behavior of ocean systems, particularly the coupled nonlinear interactions of fluids with generic structural components and unmanned platforms. Development is primarily in: detection and imaging techniques applicable to underwater and buried objects; techniques for neutralization of explosive devices; underwater life support technologies; surface and subsurface transport systems for Navy Special Forces; technologies for coastal zone mobile autonomous platform systems; and capabilities for rapid clearance of mines and obstacles from the surf zone.

Tactical Sensing

Reconnaissance

Mine reconnaissance is concerned with determining the presence of mines and obstacle fields and their densities and boundaries. The current focus of funding is on clandestine reconnaissance to enable data collection in the battlefield while preventing enemy forces from deciphering naval strategies prior to an amphibious assault. Acoustic minefield reconnaissance, would benefit from the use of underwater vehicles carrying appropriate minehunting sensors. Sensor and signal-processing approaches that support and/or facilitate high-area minefield search rates while providing sufficient clutter rejection in the coastal zone is of high priority, particularly in shallow water and very shallow water. Other efforts of high priority include synthetic aperture sonar, side-looking sonars, ahead-looking sonars, and algorithms to automatically detect, classify, and identify individual mines.

Minehunting

Current minehunting systems are time consuming, as they operate at a lower resolution than mine detection sensors. Thus, a significant amount of time is spent positioning near the target. Consequently, a high-priority research effort is the development of sensor and signal-processing algorithms to provide classification at ranges comparable to those of detection.

A new principle of minehunting being developed capitalizes on recent breakthroughs in acoustic modem communication technologies. Essentially, a large number of small, cost-effective underwater vehicles operating in parallel, equipped with high-baud-rate modems, are deployed in a region. A network of fixed acoustic buoys or nodes controls their dispersal, and the position of each candidate mine-like object is transmitted to the vehicles. An underwater positioning system, controlled by the buoys or nodes, acts similarly to Loran C, so that no transmissions from the vehicles to the nodes are required for positioning. Re-acquisition of targets is minimized, and the underwater vehicles determine the nature of the target. The result of this identification is then transmitted to the acoustic nodes, which relay the data to a central command and control system aboard a ship safely out of the minefield. The concept of a multiplicity of vehicles, acoustic nodes, and underwater positioning should radically reduce the time necessary to perform minehunting, and set the stage for mine neutralization. At present, ONR is actively soliciting new proposals to pursue all or part of this system, particularly the development of sensor technologies.

Mine Neutralization

Mine neutralization is time intensive and requires putting humans at risk. In shallow water regions, remotely operated vehicles (ROVs) can be used, but the time spent deploying and recovering these systems prevents the in-stride capability desired by the Navy and Marine Corps. This ability to clear mines and obstacles in-stride remains a critical science and technology challenge that has not yielded a straightforward solution. High-priority efforts include novel approaches to neutralizing floating, moored, buried, and bottom mines in the coastal zone.

Processes and Prediction Division

The goal of the Processes and Prediction Division of OAS S&T is to ensure the collection of environmental data, develop and improve the understanding of environmental variability, and increase the limits of predictability by using quantitative models. This division plans, fosters, and encourages an extensive program of scientific research and technological development in the following fields of interest to mine warfare.

Environmental Optics

The aim of the Environmental Optics Program is to further the understanding of how light interacts with the water column, ocean boundaries, and the atmosphere near the ocean surface. Specific areas of study include radiative transfer modeling, instrument development, optical process studies, and coastal remote-sensing.

The products of these research areas support the development and application of ocean prediction models, new ocean remote-sensing systems, and associated image analysis algorithms. Applied research in support of mine warfare and special operations is funded in areas of underwater imaging and hyperspectral remote-sensing.

Physical Oceanography

The objective of this research program is to support process-oriented, hypothesis-driven physical oceanographic research to provide tools and techniques that can be adapted for fleet use. In response to post-Cold War naval strategy and tactics, increased operational importance is being given to the coastal zone. Consequently, research emphasis is on the integration of circulation theory and modeling in nearshore environments. There is also a strong emphasis on interdisciplinary research. The primary objective of the program is to foster the transition of research products, such as numerical and theoretical models, analysis algorithms, in situ data, and seagoing instrumentation, into operational naval systems.

Biological and Chemical Oceanography

The goal of the Biological Oceanography Program is to provide information on the distribution, growth, and abundance of biota in the coastal ocean and sediments by understanding how biota influence operationally important optical and acoustical properties. An important aim of the Biological Oceanography Program is to improve observational capabilities at small-to-meso temporal and spatial scales (minutes-weeks, cm-km) and enhance the understanding of processes affecting biological phenomena. These capabilities will enable the development of new instrumentation to sample and observe biological processes and will aid in modeling coupled biophysical and bio-optical phenomena in the coastal ocean and shallow water sediments.

The aims of the chemical oceanography program are to develop predictive capabilities for chemical distributions and speciation in marine environments, especially as they relate to optical and acoustical properties of seawater; to monitor and understand chemical reactions impacting environmental quality; and to develop new in situ chemical sensors for accurate and rapid detection of key chemical species at low concentrations. Specific research projects include investigations on the occurrence and production of colored dissolved organic matter in the coastal ocean, air-sea gas exchange processes, aerosol chemical dynamics, chemistry of trace elements in the upper ocean, nutrient dynamics, and heavy metal speciation in sediments.

Ocean Modeling and Prediction

This research program seeks to develop accurate temporal and spatial quantitative models of ocean systems. Underlying fundamentals include ocean field estimation; scale and boundary interactions applied toward groundtruthing and forecast skills; subgrid-scale parameterization; and development of models incorporating ocean-atmosphere, ocean-bottom coupling, and nested domains. The goal of enhanced predictability is achieved through research on better dynamic formulations, improved numerical methods, and optimal data assimilation through adaptive sampling. Basic and applied research are pursued jointly to improve strategic and tactical decisions using environmental information and to motivate new understanding by operational experience.

Marine Geology and Geophysics

The Marine Geology and Geophysics Program, in line with changing naval operations, has increased its interest in nearshore studies and is undertaking substantive efforts in continental shelf research. One such effort is the Strata Formation on Margins (STRATAFORM) Program, which seeks to understand the creation of the continental shelf and slope stratigraphic record as a product of geological processes acting with spatial and temporal heterogeneity. This understanding will lead to the development of models capable of predicting stratigraphic patterns on a variety of continental margins. Overall, the program goal is to increase the understanding of mechanisms controlling the structure, history, and dynamics of geologic features, which in turn affect sound propagation, seafloor instability, bathymetry, and electromagnetic transmissions in the water column and ocean bottom.

Marine Meteorology and Atmospheric Effects

The objective of this research program is to sponsor integrated basic, applied, and developmental research to improve the modeling and prediction of meteorological parameters critical to naval platform, sensor, and weapons performance. The program includes research and development leading to enhanced environmental support for operations, training, mission planning, and systems development. Specific research interests include investigations of:

- mesoscale coastal phenomena, including data assimilation protocols incorporating high-data rate, asynchronous sensors (radar, Light Detection and Ranging [LIDAR], etc.);
- global, mesoscale, and on-scene modeling that focuses on the marine atmosphere and coastal zone;
- simulations to visualize weather phenomena; marine boundary layer processes, including aerosols and clouds; and
- environmental effects on electromagnetic and electro-optic propagation, and tropical cyclone behavior, including the evolution of motion and structure.

Mine Countermeasures Program Office

The Mine Countermeasure Program Office is also in the OAS S&T Department. This office is the focus of the Navy's science and technology efforts with regard to MCM. There are currently two main initiatives in this program office: the Joint Countermine Advanced Concept Technology Demonstration (ACTD) and the emerging Enhanced Naval Capability Initiative in Very Shallow Water Organic Mine Countermeasures. The objective of these projects is to develop seamless sea-to-land amphibious MCM operations.

APPENDIX F

Office of Naval Research Programs to Encourage Research in Areas of Interest to Naval Operations

Further information on the programs listed below can be found on the ONR Web Page (http://www.onr.navy.mil)

SECRETARY OF THE NAVY/CHIEF OF NAVAL OPERATIONS OCEANOGRAPHIC RESEARCH CHAIR AND THE ONR/INSTITUTION SCHOLAR

The Office of Naval Research accepts proposals for the Secretary of the Navy/Chief of Naval Operations Oceanographic Research Chair and the ONR/Institution Scholar. The awards recognize distinguished academic ocean scientists and were created to further recognize that oceanography is a core Navy competency. In addition to promoting excellent research and education, the program fosters closer relationships between the ocean science community and the operating Navy. The award includes four years of support for a Research Chair and associated ONR/Institution Scholar, and at least two graduate students.

Oceanographic Research Chairs and ONR/Institution Scholars develop scientific collaborations with other Navy and Marine Corps activities; advise ONR on policy and procedures for support of high-quality oceanography science and technology; participate in ONR Department Reviews; help identify promising science and technology opportunities in oceanographic sciences; and participate in the synthesis of recently completed ONR initiatives. Additionally, Chairs and Scholars may be asked to represent the Navy and ONR science and technology efforts to the Fleet, other agencies, and general audiences. The Chairs serve as mentors to the Scholars and the graduate students who will be conducting Navy-relevant research.

Detailed information about eligibility, qualifications and duties of the SECNAV/CNO Chair and ONR/Institutional Scholar, proposal format guidance, proposal selection criteria, and instructions for submitting proposals may be obtained on the ONR Homepage on the World Wide Web site, http://www.onr.navy.mil or by mail from the Office of Naval Research, Code 32, Attn: FY-99 Chairs Info, 800 N. Quincy Street, room 407, Arlington, VA 22217-5660.

NAVAL SCIENCE ASSISTANCE PROGRAM (NSAP)

The Naval Science Assistance Program (NSAP) serves as a two-way bridge between the warfighter and the technical community. NSAP provides on-the-spot technical assistance to joint Navy, and Marine Corps operational commands worldwide through a cadre of carefully selected, experienced Science and Technology (S&T) Advisors.

With this strong team of S&T Advisors (STAs), NSAP solves real operational problems in a rapid and inexpensive manner through the evaluation and insertion of mature technologies. NSAP facilitates the identification of operational readiness deficiencies in order to influence longer term S&T investments. Additionally, NSAP produces an annual compendium of mature technologies (not yet in the acquisition portfolio), the Technologies for Rapid Response (Blue Book), that are available to Fleet Commanders for early at-sea evaluation. Over the past few years, over 100 products have been developed and inserted into the Fleets/Forces as a result of NSAP's efforts.

NSAP offers opportunities to civilian scientists and engineers to become a member of the NSAP team that includes the Program Office at ONR, S&T Advisors (STAs) located at commands throughout the world, and Program Managers at participating naval laboratories and warfare centers. NSAP assignments vary in length and recruitment for vacancies is done annually. A comprehensive training program is provided for all new STAs prior to each assignment. This training focuses on technical and personal skills to accommodate job requirements and relocation needs.

NSAP is at the tip of the S&T spear, as both an ONR Code (N09N) and as a CNO N091 Code (N911N). The purpose of this dual role is to provide a real and visible commitment on the part of both ONR and N091 to work collaboratively with the Fleet/Forces, to help them more clearly define their technology needs, and to accelerate technology transitions to them. Forming this collaborative partnership between the warfighter and the S&T community results in:

better identification of needed capabilities,
expert evaluation of emerging technologies that might deliver those capabilities,
rapid affordable solutions, and
a stronger influence on S&T and acquisition investments.

Program Points of Contact
Susan L. Bales, Director
(703) 696-2901
nsapdir@nosc.mil

George F. Stimak,
Deputy Director for Technology
(703) 696-2902
nsapdept@nosc.mil

David R. Starkston,
Deputy Director for Operations
(703) 696-4018
nsapdepo@nosc.mil

POSTDOCTORAL FELLOWSHIP PROGRAM

The Office of Naval Research (ONR) sponsors a Postdoctoral Fellowship Program for appointments at a number of naval R&D centers and laboratories. The objective of this program is to encourage the involvement of creative, capable and highly trained scientists and engineers who have received a Ph.D. or equivalent within the prior seven years in areas of great interest and relevance to the Navy.

Participants selected for support under the ONR Postdoctoral Fellowship Program must be United States citizens or nationals. Participants must be eligible for a Department of Defense security clearance of "Secret." In most cases, participants will be permitted to initiate research pending completion of the security clearance. All appointments are contingent upon Fellows obtaining and maintaining the appropriate level of security clearance.

Fellowship awards will be based upon the technical quality and relevance of the proposed research, recommendations by the Navy laboratory/ center, academic qualifications, letters of reference, and availability of funds. Selection of Fellows occurs four times per year.

Participating Navy laboratories and centers include:
Naval Air Warfare Center (Warminster, PA, China Lake, CA, Orlando, FL)
Naval Command, Control and Ocean Surveillance Center (San Diego, CA)
Naval Facilities Engineering Service Center (Port Hueneme, CA)
Naval Research Laboratory (Washington, DC, Stennis Space Center, MS, Monterey, CA)
Naval Surface Warfare Center/Carderock (Bethesda, Annapolis, White Oak, MD)
Naval Surface Warfare Center/Dahlgren (Dahlgren, VA, Panama City, FL, White Oak, MD)
Naval Undersea Warfare Center/Newport (Newport, RI and New London, CT)
Naval Medical Research and Development Command (Bethesda, MD)
The Naval Medical Research Institute (NMRI) (Bethesda, MD)
NMRI's Toxicology Detachment (Wright Patterson Air Base, Dayton, OH)
The Naval Health Research Center (San Diego, CA)
The Naval Aerospace Medical Research Laboratory (Pensacola, FL)
The Naval Biodynamics Laboratory (New Orleans, LA)
The Naval Submarine Medical Research Laboratory (Groton, CT)
The Naval Dental Research Institute (Great Lakes, IL)
Navy Personnel Research and Development Center (San Diego, CA)
United States Naval Academy (Annapolis, MD)

Contractor Contact
American Society for Engineering Education
Projects Office
1818 N Street, N.W., Suite 600
Washington, DC 20036-2479
Phone: (202) 331-3525
Email: projects@asee.org
Web Site: http://www.asee.org/fellowships/html/onr.htm

ONR Point of Contact
(For policy issues only)
Name: Anthony Junior
Title: Program Manager
Phone: (703) 696-0453
Office: Corporate Programs Division
Email: juniora@onr.navy.mil

Address
Office of Naval Research
ATTN: ONR 363, Room 804
800 N. Quincy Street
Arlington, VA 22217-5660

YOUNG INVESTIGATOR PROGRAM

The objectives of the Young Investigator Program (YIP) are to attract to naval research outstanding new university faculty members, to support their research, and to encourage their teaching and research careers. ONR anticipates making at least 18 new awards each year to academic researchers who have recently received Ph.D. or equivalent degrees.

Awards of $100,000 per year for three years, with the possibility of additional support for equipment or collaborative research with a Navy laboratory, are made, based on research proposals and supporting materials. Special attention will be given to proposals in naval priority research areas, described in the Science & Technology section of the ONR Website. In addition, it is anticipated that two of these applicants receiving an ONR Young Investigator award will also be selected to receive a Presidential Early Career Award for Scientists and Engineers.

This program is open to U.S. citizens, nationals, and permanent residents holding tenure track or permanent faculty positions at U.S. universities and colleges who have received graduate degrees (Ph.D. or equivalent) on or after the date published in the Complete Announcement.

Office of Naval Research makes a new announcement of this program each year that it is in force. Potential applicants should refer to the Complete Announcement for rules and deadlines governing the current competition.

This competition has been severe in past years. Typically, fewer than 10% of the submitted proposals have resulted in YIP awards. Past awardees have both submitted outstanding research proposals and possessed outstanding records of prior professional accomplishments. Given that "past performance" is a selection criterion, applicants are advised that the biographical information submitted as part of the proposal should list all relevant past activities.

Those proposals not selected for Young Investigator awards are automatically considered for ONR's regular research grant program in competition with all other research proposals submitted to ONR. Typically, additional proposals originally submitted to the Young Investigator Program have been selected each year for funding via the regular research grant program. Thus, the Young Investigator Program is not a "research initiation" opportunity with standards that are less demanding than ONR's regular research grant program. ONR's Young Investigator awards are intended to confer honor upon awardees beyond the research funding being provided.

Selection and Evaluation

Applicants are selected by the Office of Naval Research based on:

1. past performance, demonstrated by the significance and impact of previous research, publications, professional activities, awards and other recognition, etc;
2. a creative proposal, demonstrating the potential for making progress in an important, naval-relevant scientific area; and
3. a long-term commitment by the university to the applicant and the research.

ONR Point of Contact

See Complete Announcement for the contact for each discipline.

(For policy issues only)
Dr. Donald E. Polk
(703) 696-4111
Director
Corporate Programs Division, ONR
Email: polkd@onr.navy.mil

Address

Office of Naval Research
ATTN: ONR 363, Room 804
800 N. Quincy Street
Arlington, VA 22217-5660

SUMMER FACULTY RESEARCH PROGRAM

Program Description

The Summer Faculty Research Program provides science and engineering faculty members from institutions of higher education the opportunity to participate in research at Navy laboratories for a ten-week period during the summer break. Participants work with professional peers in the Navy laboratories on research tasks of mutual interest. The research for the summer is defined in advance through correspondence and an optional pre-program visit to the research site, for which funds are provided.

This program offers college and university faculty members an experience that can benefit them professionally. Participants have an opportunity to establish continuing research relations with the R&D personnel of the host laboratories which may result in sponsorship of the participants research at their home institutions. Professional contacts are expanded. Laboratory, computational, and specialized library facilities on a scale not available at some universities and colleges can be used to conduct research which might otherwise be impossible. Faculty from Historically Black Colleges and Universities and Minority Institutions (HBCU/MI) are especially encouraged to apply.

Participating in the Summer Faculty Research Program are:

Naval Air Warfare Center, Aircraft Division, Warminster, PA
Naval Air Warfare Center, Training Systems Division, Orlando, FL
Naval Air Warfare Center, Weapons Division, China Lake, CA
Naval Command, Control and Ocean Surveillance Center, RDT&E Division, San Diego, CA
Naval Facilities Engineering Service Center, Port Hueneme, CA
Naval Research Laboratory, Washington, DC, and Stennis Space Center, MS
Naval Surface Warfare Center, Carderock Division, Carderock and Annapolis,MD
Naval Surface Warfare Center, Dahlgren Division, Coastal Systems Station, Panama City, FL
Naval Undersea Warfare Center, Newport, RI, and New London, CT
Defense Equal Opportunity Management Institute, Cocoa Beach, FL
Navy Personnel Research & Development Center, San Diego, CA
Naval Aerospace Medical Research Laboratory, Pensacola, FL
Naval Biodynamics Laboratory, New Orleans, LA
Naval Health Research Center, San Diego, CA
Naval Medical Research Institute, Bethesda, MD
Naval Submarine Medical Research Laboratory, Groton, CT

Levels of Appointment - Three levels of appointment are available: Summer Faculty Fellow, Senior Summer Faculty Fellow, and Distinguished Summer Faculty Fellow. For all appointments, the applicant must be a U.S. citizen and hold a teaching or research position at a U.S. college or university. Levels of appointment will be determined by a committee of scientists and engineers. Decisions of the committee will be final.

Terms - Stipends vary by level of appointment and are set each year. In addition, each Fellow will be reimbursed for his/her personal travel for an optional pre-program visit to the research site. For applicants who relocate their residence for the ten-week period of appointment, a similar travel and modest relocation allowance will be provided.

Security Requirement - Although it is anticipated that the research undertaken will usually be unclassified, participants must be U.S. citizens either holding, or eligible for, a Department of Defense Security Clearance of SECRET in order to be admitted to the Navy Laboratories and Research Centers at which the program is conducted.
Period of Appointment - Ten continuous weeks on site during May, June, July and August.

Housing - Housing availability and cost vary from center to center. Navy host personnel may be able to assist participants in finding suitable housing. The pre-program visit provides a good opportunity to seek housing.

Contractor Contact
American Society for Engineering Education (ASEE)
Projects Office
1818 N Street N.W., Suite 600
Washington, DC 20036-2479
(202) 331-3525
Email: projects@asee.org
Website: *http://www.asee.org/fellowships/html/nsfp.htm*

ONR Point of Contact
(For Policy Questions Only)
Anthony Junior
(703) 696-0453
Title: Program Manager
Office: Corporate Programs
Email: juniora@onr.navy.mil

Address
Office of Naval Research
ATTN: ONR 363, Room 804
800 North Quincy Street
Arlington, VA 22217-5660

FACULTY SABBATICAL LEAVE PROGRAM

The Faculty Sabbatical Leave Program provides science and engineering faculty the opportunity to conduct research at Navy laboratories while on sabbatical leave.

Terms of Appointment - Participants in the Sabbatical Leave Program will receive a monthly stipend making up the difference between salary and sabbatical leave pay from their home institution. In addition, participants will receive reimbursement for travel to and from the laboratory site and a relocation allowance for those who must relocate their residence during their sabbatical leave tenure. Sabbatical leave appointments are contingent on the availability of sabbatical leave positions. Appointments are for a minimum of one semester and a maximum of one year. This is a residential sabbatical, thus participants must conduct research on site.

Applications - Faculty should submit applications at least six months prior to their proposed sabbatical leave starting date. Applications will be reviewed and the applicant notified of outcome within 90 days of receipt of application. ASEE will verify support of the proposed project with the sponsoring Navy Laboratory.

A complete Sabbatical Leave application includes the following:

(1) completed application form;
(2) description of proposed research project (approximately three typewritten pages); and
(3) statement from academic institution verifying the faculty member's eligibility and approval for sabbatical leave including dates.

Prospective applicants should contact the research facility at which they are interested in working to jointly develop a suitable research proposal. Also required is an estimate of base salary and percentage of salary provided by the home institution while on sabbatical leave.

Contractor Contact
American Society for Engineering Education
Projects Office, Suite 600
1818 N Street, N.W.
Washington, DC 20036-2479
(202) 331-3525
Email: projects@asee.org
Website: *http://www.asee.org/fellowships/html/nsfp.htm*

ONR Point of Contact
(For policy issues only)
Anthony Junior
(703) 696-0453
Title: Program Manager
Office: Corporate Programs Division
Email: juniora@onr.navy.mil

Address
Office of Naval Research
ATTN: ONR 363, Room 804
800 N. Quincy Street
Arlington, VA 22217-5660

MULTI-DISCIPLINARY RESEARCH PROGRAM OF THE URI (MURI)

The Multi-disciplinary Research Program of the University Research Initiative (MURI) is a multi-agency DoD program that supports research teams whose efforts intersect more than one traditional science and engineering discipline. Multi-disciplinary team effort can accelerate research progress in areas particularly suited to this approach. Multi-disciplinary research also can help to hasten the transition of research findings to practical application.

MURI awards are made in research topics specified by the participating defense agencies each year that the program is in force. Specified topics change each year. Awards are typically for a period of three years (funded incrementally or as options) with two additional years possible as options to bring the total award to five years, and at a funding level ranging from half a million to a million dollars per year, with the size of the award dependent upon the topic, technical goals, and availability of appropriations.

The MURI is competed in specific research topics described in the current announcement. Authors of potential proposals are advised to read the announcement carefully. It explains the DoD's research needs upon which the topics are based. Proposals may be submitted only by U. S. institutions of higher education (other than federal government organizations) with degree granting programs in science or engineering, or by consortia of such institutions (universities). Proposals from consortia of academic institutions may be warranted because research in the multi-disciplinary topics may require forming teams with strengths in multiple science and engineering fields. One academic institution will be the primary awardee for purposes of award execution.

The DoD expects that MURI programs will promote application of defense research, principally for defense purposes but also for commercial purposes. The research topics described in the MURI announcement generally underpin dual use defense technologies that are critical to national defense and that also have good potential for commercial application. Interactions with research and development organizations that transition research findings to application, particularly industrial organizations, DoD laboratories, and other organizations that perform research and development for defense applications, are encouraged.

Program Point of Contact
Please see the complete announcements for the technical contact for each topic.

ONR Point of Contact
(For policy issues only)
Dr. Clifford Lau
(703) 696-0431
Title: Program Officer
Office: Corporate Programs Division
lauc@onr.navy.mil

Address
Office of Naval Research
ATTN: ONR 363, Room 804
800 N. Quincy Street
Arlington, VA 22217-5660

DEFENSE UNIVERSITY RESEARCH INSTRUMENTATION PROGRAM (DURIP)

DURIP is a multi-agency DoD program within the University Research Initiative designed to improve the capabilities of U.S. institutions of higher education to conduct research and to educate scientists and engineers in areas important to national defense by providing funds for the acquisition of research equipment.

Each year that DURIP is in effect, a complete announcement will provide detailed program information, including the deadline date for submitting proposals. DURIP proposals submitted to ONR should facilitate research in an area of interest to ONR, as described in the Science and Technology section of the ONR Website. Authors of potential proposals may contact the appropriate program managers, listed with each area of interest, to explore possible mutual interests before submitting proposals.

DURIP funds will be used for the acquisition of major equipment to augment current or develop new research capabilities in support of DoD-relevant research. Proposals may request $50,000 to $1,000,000. Cost sharing is encouraged. Proposals for purely instructional equipment are not eligible. General-purpose computing facilities are not appropriate for DURIP funding, but requests for computers for DoD-relevant research programs are appropriate. Funds under DURIP will not be used for construction, including building or facilities modification.

Lastly, costs (e.g., machine shop expenses) for constructing, assembling, and/or installing equipment may be proposed, but costs for continued operation and maintenance are not eligible for consideration. This competition is open to U.S. institutions of higher education, other than those that are a federal government organization, with degree granting programs in science, math, or engineering.

Proposals are evaluated competitively. The evaluation criteria are listed in the current program announcement.

ONR Point of Contact
See Science and Technology section on this page to identify the program manager for each ONR research area of interest.

(For policy issues only)
Dr. Donald Polk
Phone: (703) 696-4111
Title: Director
Office: Corporate ProgramsDivision
Email: polkd@onr.navy.mil

Address
Office of Naval Research
ATTN: ONR 363, Room 804
800 N. Quincy Street
Arlington, VA 22217-5660

DEFENSE EXPERIMENTAL PROGRAM TO STIMULATE COMPETITIVE RESEARCH

The Office of Naval Research participates in the Department of Defense Experimental Program to Stimulate Competitive Research (DEPSCoR), within the University Research Initiative (URI). The program is conducted with the cooperation of the Experimental Program to Stimulate Competitive Research (EPSCoR) State Committees.

DEPSCoR is in response to Congressional requirements to stimulate competitive research in states that have not traditionally been recipients of a large amount of Federal research awards. DEPSCoR is designed to improve the capabilities of U.S. institutions of higher education ("universities") in eligible states (see Complete Announcement), to conduct research and to educate scientists and engineers in areas important to national defense. Only U.S. institutions of higher education with degree granting programs in science, mathematics, or engineering are eligible for DEPSCoR grants. Proposals from eligible universities must be submitted via the EPSCoR State Committee. Proposals not submitted through an EPSCoR State Committee will be rejected.

Proposals to perform research in technical areas identified as being of interest to ONR, or other areas important to national defense, will be considered. Potential proposers are advised to consult the Science and Technology section of ONR's home page and to contact ONR program managers to explore possible mutual interests before submitting proposals. In this way, DEPSCoR contributes to the states' goals of developing new research capabilities while simultaneously supporting DoD research goals.

DEPSCoR funds may be used to develop new research capabilities in support of DoD research goals. Proposals may contain requests for equipment necessary for the completion of the proposed research. General-purpose computing facilities and purely instructional equipment are not appropriate for DEPSCoR funding, but requests for funding for specialized computers entirely devoted to specific DEPSCoR research programs are appropriate. Funds under DEPSCoR may not be used for buildings or facilities modification.

Proposals are evaluated competitively. The evaluation criteria are listed in the current program announcement.

Program Point of Contact
See Science and Technology section on this page to identify the program manager for each ONR research area of interest.

(For Policy Issues Only)
Name: Dr. Donald E. Polk
Phone: (703) 696-4111
Director, Corporate Programs Division

Address
Office of Naval Research
ATTN: ONR 363, Room 804
800 North Quincy Street
Arlington, VA 22217-5660

APPENDIX G

Office of Naval Research Program Contact Information

Ocean, Atmosphere, and Space Science Technology (OAS S&T) Department

Overall Department

General Points of Contact:

Dr. Steven E. Ramberg, Department Head
Phone: (703) 696-4358
Email: rambers@onr.navy.mil

Captain Tim Schnoor, USN, Deputy
Phone: (703) 696-5123
Email: schnoor@onr.navy.mil

Focus Area Points of Contact:

Battlespace Environments (BSE)
Dr. Melbourne Briscoe
Phone: (703) 696-4120
Email: briscom@onr.navy.mil

Anti-Submarine Warfare (ASW)
Dr. Frank Herr (acting)
Phone: (703) 696-4125
Email: herrf@onr.navy.mil

Mine Warfare (MIW)
Mr. Doug Todoroff
Phone: (703) 696-2485
Email: todorod@onr.navy.mil

Maritime Intelligence, Surveillance, and Reconnaissance and Space Exploitation (ISR)
Dr. Frank Herr
Phone: (703) 696-4125
Email: herrf@onr.navy.mil

Countermine Program Office Points of Contact:

Captain Tim Schnoor, USN, Program Manager
Phone: (703) 696-5123
Email: schnoor@onr.navy.mil

Mr Doug Todoroff, Program Manager, MCM Enhanced Naval Capability Initiative
Phone: (703) 696-2485
Email: todorod@onr.navy.mil

Department Divisions

Sensing and Systems Division Points of Contact:

Dr. Frank L. Herr, Director
Phone: (703) 696-4125
Email: herrf@onr.navy.mil

CDR Scott Tilden, Deputy
Phone: (703) 696-8698
Email: tildens@onr.navy.mil

Ocean Acoustics Program:

Dr. Jeffrey Simmen, Team Leader
Phone: (703) 696-4204
Email: simmenj@onr.navy.mil

Dr. Ellen Livingston, Program Officer
Phone: (703) 696-4203
Email: livinge@onr.navy.mil

Dr. Edward Estalote, Program Officer
Phone: (703) 696-6940
Email: estaloe@onr.navy.mil

Ms. Beverly Kuhn, Program Specialist
Phone: (703) 696-6998
Email: kuhnb@onr.navy.mil

Remote-sensing and Space Program:

Dr. Robert McCoy
Phone: (703) 696-8699
Email: mccoyr@onr.navy.mil

Dr. Dennis Trizna
Phone: (703) 696-8380
Email: triznad@onr.navy.mil

Mr. Charles Luther
Phone: (703) 696-4123
Email: lutherc@onr.navy.mil

Dr. Marie C. Colton
Phone: (703) 588-0705
Email: coltonm@onr.navy.mil

Sensing – Information Dominance Program:

Dr. David Johnson
Phone: (703) 696-0807
Email: johnsodh@onr.navy.mil

Coastal Dynamics Program:

Dr. Thomas Kinder, Team Leader
Phone: (703) 696-1206
Email: kindert@onr.navy.mil

Dr. Linwood Vincent, Army Liaison
Phone: (703) 696-4118
Email: vincenc@onr.navy.mil

Dr. Reginald Beach, Program Officer
Phone: (703) 696-6523
Email: beachr@onr.navy.mil

Sensors, Sources, and Arrays Program:

Dr. Ken Dial
Phone: (703) 696-0806
Email: dialk@onr.navy.mil

Ocean Engineering and Marine Systems Program:

Dr. Tom Swean, Team Leader
Phone: (703) 696-4025
Email: sweant@onr.navy.m

Undersea Signal Processing Program:

Ms. Nancy Harned, Team Leader and Active Sonar Signal Processing Program Manager
Phone: (703) 696-4758
Email: harnedn@onr.navy.mil

Dr. John Tague, Passive Sonar Signal Processing and Fundamental Research Initiatives Program Manager
Phone: (703) 696-4399
Email: taguej@onr.navy.mil

Tactical Sensing Support Program:

Dr. Randall Jacobson
Phone: (703) 696-4121
Email: jacobsr@onr.navy.mil

Processes and Prediction Division
Points of Contact:

Dr. Melbourne G. Briscoe, Director
Phone: (703) 696-4120
Email: briscom@onr.navy.mil

CDR Fred Pfeil, Deputy
Phone: (703) 696-4395
Email: pfeilf@onr.navy.mil

Environmental Optics Program:

Dr. Steven Ackleson, Program Officer
Phone: (703) 696-4732
Email: ackless@onr.navy.mil

Dr. Joan Cleveland, Program Officer
Phone: (703) 696-4532
Email: clevelj@onr.navy.mil

Physical Oceanography Program:

Dr. Louis Goodman, Team Leader
Phone: (703) 696-4112
Email: goodmal@onr.navy.mil

Dr. Stephen P. Murray, Program Officer
Phone: (703) 696-4533
Email: murrays@onr.navy.mil

Biological/Chemical Oceanography Program:

Dr. James Eckman, Team Leader (Biological and Chemical Oceanography)
Phone: (703) 696-4590
Email: eckmanj@onr.navy.mil

Dr. Ron Tipper, Program, Officer (Biological Oceanography)
Phone: (703) 696-4415
Email: tipperr@onr.navy.mil

Dr. Ron Ferek, Program, Officer (Chemical Oceanography)
Phone: (703) 696-0518
Email: ferekr@onr.navy.mil

Dr. Joan Cleveland, Program Officer (Biological Oceanography andEnvironmental Optics)
Phone: (703) 696-4532
Email: clevelj@onr.navy.mil

Ocean Modeling and Prediction Program:

Dr. Thomas B. Curtin, Team Leader
Phone: (703) 696-4119
Email: curtint@onr.navy.mil

Dr. Manuel (Manny) Fiadeiro, Scientific Officer
Phone: (703) 696-4441
Email: fiadeim@onr.navy.mil

Marine Geology and Geophysics Program:

Dr. Joseph H. Kravitz, Program Officer
Phone: (703) 696-4070
Email: kravitj@onr.navy.mil

High Latitude Dynamics Program:

Dr. Dennis Conlon
Phone: (703) 696-4720
Email: conlond@onr.navy.mil

Marine Meterology and Atmospheric Effects Program:

Dr. Scott Sandgathe, Team Leader
Phone: (703) 696-0802
Email: sandgas@onr.navy.mil

Dr. Robert Abbey
Phone: (703) 696-6598
Email: abbeyr@onr.navy.mil

Dr. Ronald Ferek
Phone: (703) 696-0518
Email: ferekr@onr.navy.mil

Dr. James Edson
Phone: (703) 696-2495
Email: edsonj@onr.navy.mil

Appendix H

Acronyms and Abbreviations

3-D	Three-Dimensional
3GWAM	Third Generation Wave Model
ACTD	Advanced Concept Technology Demonstration
ADCP	Acoustic Doppler Current Profiler
ALMDS	Airborne Laser Mine Detection System
AMCM	Airborne Mine CounterMeasures
AMNS	Airborne Mine Neutralization System
ASCII	American Standard Code for Information Interchange
ASW	Anti-submarine warfare
AUV	Autonomous Vehicle
BSP	Battlespace Profiler
BUFR	Binary Universal Form of Representation
C^2I	Command, Control, and Information
C^4I	Command and Control, Communications, Computers and Intelligence
CCM	Counter-Counter Measures
CDOM	Chromophoric Dissolved Organic Material
CEAS	Comprehensive Environmental Assessment System
CLZ	Craft Landing Zone
CNMOC	Commander, Naval Meteorology and Oceanography Command
CNO	Chief of Naval Operations
COAMPS	Coupled Ocean/Atmosphere Mesoscale Prediction System
CODAS	Common Oceanographic Data Archival System
COE	Common Operating Environment
COMINEWARCOM	Commander, Mine Warfare Command
COMOMAG	Mobile Mine Assembly Group
CONOPS	Concept of Operations

CoOP	Coastal Ocean Processes
COTS	Commericial Off-The-Shelf
CTD	Conductivity, Temperature, Depth
DBDB	Digital Bathymetric Data Base
DET	Distributed Explosive Technology
DII COE	Defense Information Infrastructure Common Operating Environment
DPSR	Distributed Processing System Replacement
DW	Data Warehouse
EIP	Explode-In-Place
ENATD	Enhanced Neutralization Advanced Technology Demonstration
ENVIRON MON	Environmental Monitoring
EOD	Explosive Ordinance Disposal
ERDAS	Earth Resources Data Acquisition System
FGDC	Federal Geographic Data Committee
FLENUMMETOCCEN	Fleet Numerical Meterology and Oceanography Center
FTP	File Transfer Protocol
GCCS – M	Global Command and Control System – Maritime
GDB	Geophysical Data Base
GDEM	Generalized Digital Environmental Model
GIF	Graphics Interchange Format
GIS	Geographic Information System
GPS	Global Positioning System
GRIB	Gridded Binary
GRP	Glass-Reinforced Plastic
GSF	Generic Sensor Fomat
GUI	Graphic User Interface
HIDEX	High Intake Defined Excitation bathyphotometer
HIS	Interactive Histogram Stretching
HM 15	Helicopter Hanger
HTML	Hyper Text Markup Language
HUNT	Minehunting
HYSAS	Hydrographic Source Assessment System
ISBHM	Intelligent Self-Burying Hunter Mine
JTA	Joint Technical Architecture
LCID	Littoral Currents Information Data Base
LIDAR	Light Detection and Ranging
LMRS	Long-term Mine Reconnaissance System
M&S	Modeling and Simulation
MACAS	Magnetic Capability and Safety
MCM	Mine Countermeasures
MCM REP-UWCOND	MCM Report – Underwater Conditions

MCMRON	Mine Countermeasures Squadron
MCOG	Mine Warfare Coordinating Group
MCS	Mine Countermeasures Command, Control and Support Ship
MEDAL	Mine Warfare Environmental Decision Aid Library
MEDEA	Measurements of Earth Data for Environmental Analysis
METOC	Meteorological and Oceanographic
MH-53E	Minehunting Helicopter, 53E version
MHC	Mine-hunter, Coastal ship
MICFAC	Mobile Integrated Command Facility
MIREM	Mine Warfare Readiness and Effectiveness Measurements
MIW	Mine Warfare
MLDC	Magic Lantern Deployment Contingency
MMS	Marine Mammal System
MNS	Mine Neutralizaton System
MOMAU	Mobile Mine Assembly Unit
MOODS	Master Oceanographic Observation Dataset
MSDDB	Master Seafloor Digital Data Base
MSO	Minesweeper, Ocean Going
MTEDS	MCM Tactical Environmental Data System
MWP	Mine Warfare Pilot
MWTC	Mine Warfare Training Center
N34D	Data Ingest Staff
N34	Ocean/Seafloor Data Division
N342	SWATH Systems Branch
N344	Validation Branch
N35	Geophysics and Acoustics Data Division
N351	Geophysics Branch
N5	Geospatial Databases Department
N51	Oceanographic Databases Division
N52	Seafloor Databases Division
N54	Applied Hydrography Division
N931	Laboratory Branch
N95	Modeling and Techniques Division
N951	Simulation and Visualizaton Branch
N85	Director, Expeditionary Warfare
NAS	Naval Air Station
NATO	North Atlantic Treaty Organization
NAVOCEANO	Naval Oceanographic Office
NAVSEA	Naval Sea Systems Command
NES	Network Encryption System
NEUT	Neutralization
NIDAS	Naval Interactive Data Analysis System
NIPRNET	Nonsecure Internet Protocol Route Network
NITES	Navy Integrated Tactical Environmental Subsystem
NMRS	Near-term Mine Reconaissance System
NODC	National Oceanographic Data Center
NOGAPS	Navy Operational Global Atmospheric Prediction System
NOMBO	Non-Mine, Mine-Like Bottom Object
NOO	Naval Oceanographic Office

NRC	National Research Council
NSW	Naval Special Warfare
NSWC-DD CSS	Naval Surface Warfare Center, Dahlgren Division, Coastal System Station
NTCS – A	Navy Tactical Computer System – Afloat
NWP	Naval Warfare Publication
OAS S&T	Ocean, Atmosphere and Space Science and Technology
OES	Oceanographic/Environmental Summary
ONR	Office of Naval Research
OPAREA	Operating Area
OPNAV N85	Director, Expeditionary Warfare
OPNOTE	Operational Note
OSB	Ocean Studies Board
OSS	Oceanographic Support Subsystem
OSS DW	Oceanographic Support Subsystem Data Warehouse
PC	Personal Computer
PINS	Precise Integrated Navigation System
PMA	Post Mission Analysis
PMI	Production Modernization Initiative
POM	Program Objectives Memorandum
PW	Propelled Warhead mines
QC	Quality Control
Q-MIPS	Q-route Mine Image Processing System
R&D	Research and Development
RAMICS	Rapid Airborne Mine Countermeasures System
RDBMS	Relational Data Base Management System
RDI	RD Instruments
RDP	Raw Data Point
RECO	Remote Control
RECON	Reconnaissance
RMOP	Remote Minehunting Operatonal Prototype
RMS	Remote Minehunting System
ROV	Remotely Operated Vehicle
RTDHS	Real Time Data Handling System
SABRE	Shallow Water Assault Breaching System
SCADCP	Self-Contained Acoustic Doppler Current Profiler
SCBP	Self-Contained Bioluminescence Photometer
SIG MOD	Signature Modification
SIPRNET	Secret Internet Protocol Router Network
SLMM	Submarine Launched Mobile Mine
SLOC	Sea Lanes of Communication
SMCM	Surface Mine CounterMeasures
SQL	Structured Query Language
SSN	Attack Submarine(Nuclear)
STOIC	Special Tactical Oceanographic Information Chart
STRATAFORM	Strata Formation on Margins Program

SWABS	Shallow water Assault Breaching System
SWEEP	Minesweeping
SWGDEM	Shallow water Generalized Digital Environmental Model
SZ	Surf Zone
TAMDA	Tactical Acoustics Measurement and Decision Aid
TDA	Tactical Decision Aid
TESS	Tactical Environmental Support System
TOWDEX	Towed Defined Excitation Bathyphotometer
UAV	Unmanned Aerial Vehicles
UMCM	Undersea/water Mine Countermeasures
UNISIPS	Unified Sonar Image Processing System
URI	University Research Initiative
UUV	Unmanned Undersea Vehicle
VMADCP	Vessel-Mounted Acoustic Doppler Current Profiler
VSW	Very Shallow Water
WW I	World War I
WW II	World War II
WWW	World Wide Web
XBP	Expendable Bathyphotometer
XBT	Expendable Bathythermograph
XSV	Expendable Sound Velocity